Process Industry Economics

Process Industry Economics
Principles, Concepts and Applications

Second Edition

David Brennan
Associate Professor, Department of Chemical Engineering,
Monash University, Australia

ELSEVIER

Elsevier
Radarweg 29, PO Box 211, 1000 AE Amsterdam, Netherlands
The Boulevard, Langford Lane, Kidlington, Oxford OX5 1GB, United Kingdom
50 Hampshire Street, 5th Floor, Cambridge, MA 02139, United States

Library of Congress Cataloging-in-Publication Data
A catalog record for this book is available from the Library of Congress

British Library Cataloguing-in-Publication Data
A catalogue record for this book is available from the British Library

ISBN: 978-0-12-819466-9

For information on all Elsevier publications
visit our website at https://www.elsevier.com/books-and-journals

Publisher: Susan Dennis
Acquisitions Editor: Kostas KI Marinakis
Editorial Project Manager: Sara Valentino
Production Project Manager: Joy Christel Neumarin Honest Thangiah
Cover Designer: Miles Hitchen

Typeset by SPi Global, India

Contents

Preface

When the first edition of this book was published by the Institution of Chemical Engineers in 1998, the importance of environmental protection and process safety was well established, and sustainability was emerging as an important criterion for industry performance. In 2020, the chemical engineering profession, along with industry and government, is wrestling with strategies for minimising and adapting to global warming, recycling of waste materials, and sustainable provision of resources, energy, and products. Technology options are keenly sought for development, assessment, and implementation.

The principles of process industry economics have been integral with the chemical engineering profession since its inception, and have been taught in undergraduate courses, although relatively briefly. Economics comes to life when graduates begin their professional careers. It is soon discovered for example by the young chemical engineer in an operational context, that any improvement to an operating plant requires capital investment, and incurs changes in ongoing costs and revenues; this must all be evaluated, documented, and justified to management.

In writing this second edition of the book, I have sought to trace some important developments in the learning experience leading to improvements in environmental, safety, and sustainability performance, and have explored the links between their assessment and economic evaluation. Some specific cases involving environmental and sustainability improvements in different industry contexts are explored, reflecting market, location, technology, and cost constraints commonly encountered. In approaching market evaluation, the scope has been widened to include fuel and mineral commodities, recognising their important roles as feedstocks or energy sources in terms of quality, availability, and price.

The basic principles of economic evaluation have not changed significantly over the years, and much is owed to valuable foundation work by earlier writers. Their contribution is acknowledged in exploring those principles. Regrettably, there has been an apparent decline in publicly available data on current chemical prices and production levels in different countries, and current costs of construction materials, equipment, and plants. There has been an apparent shift from public disclosure of data, to greater dependence on commercial access. This provides a challenge for education, research, and investigative work.

Expertise in market evaluation, capital cost estimation, operating cost estimation, and profitability evaluation, the foundation pillars of process industry economics, is diverse, and almost always held by different personnel with quite different levels of knowledge, skill and experience. This presents a particular challenge for chemical engineers involved in economic evaluation, where communication with specialists in these areas is critical for effective project development and economic assessment. Communication between economic evaluators and evaluators of safety, environmental and sustainability performance is similarly a major challenge for effective project development and appraisal. Finally, sound communication skills are vital in preparing expenditure proposals for approval by management, and in advising stakeholders regarding the public benefits of approved projects.

In the later chapters of this revised edition, the perspective is widened from individual projects to include the process industries at large. Aspects discussed include industrial classification, industry ownership, the role of government, and industrial integration. The potential for the chemical engineering profession to contribute to effective government policy in a number of aspects impinging on industry development is reviewed. Finally, the importance of human resources across industry, research organisations, government bodies, and educational institutions working together to provide a productive industrial economy is discussed.

I owe much of my knowledge in process economics to experience in design, operation and project evaluation gained in industry, and from teaching and research experience at university. I have also benefited from industry support in providing research support, work experience, and advice during my academic career at RMIT and Monash Universities, including research at Melbourne University for my PhD. Such interaction between academic and industrial personnel is a vital aspect of developing a 'professional ecology'.

Whilst my experience has been mostly (though not exclusively) gained in Australia, I am confident that the concepts and principles explored in the book are applicable globally, and that the applications and cases considered have a strong degree of relevance for other countries.

David Brennan

Acknowledgements

I thankfully acknowledge the contributions and guidance from the following bodies and individuals in assisting with the writing of the second edition of this book.

- The Institution of Chemical Engineers for initiating and supporting publication of the revised edition of the book
- Staff of Elsevier for support at various stages of the book's development, especially
  - Kostas Marinakis at the planning phase
  - Sara Valentino during the writing of the book
  - Indhumathi Mani for assistance with permissions
  - Honest Thangiah, Joy Christel Neumarin at the production phase
- The Department of Chemical Engineering at Monash University for supporting my endeavour
- Kerry Pratt, Andrew Hoadley, Kurt Golonka, Manjusha Thorpe, and Jennifer Davies for help in reviewing chapters
- Nick Brennan, Lilyanne Price, Trina Olcorn, and Daniel Rojas Sanchez for assistance with diagrams
- My wife Elaine, for her patience throughout the project.

David Brennan

CHAPTER

The scope of process industry economics

1

Economics encompasses all things and all people.
David Brennan

Chapter outline

1.1 Gifts and challenges of resources

Planet Earth is generously endowed with natural resources such as air, water, and soil, and a wide diversity of plants and living creatures, all making an invaluable contribution to human life. It is also endowed with a vast range of extractable fuel and mineral resources such as crude oil, natural gas, iron ore, bauxite, phosphate rock, and salt. Extractable resources from land or sea, as well as agricultural products such as wheat and sugarcane, can be progressively processed to yield a wide spectrum of more valuable products. Such products include fuels, metals, polymers, pharmaceuticals and foods, which find use in a wide range of human needs and activities. Despite their inherent potential value, almost all extractable resources contain impurities and demand refinement before further processing and use. Thus natural gas at the wellhead may contain sulphur, water, and carbon dioxide, while minerals such as lead and zinc occur as sulphides but are associated with gangue. Extractable resources are commonly located at sites remote from human population and the point of demand, posing challenges in site location for their further processing. Thus an extensive network of processing and transport is required to deliver extractable resources and their intermediate and downstream products to their ultimate point of use at the market place. This network has

Process Industry Economics. https://doi.org/10.1016/B978-0-12-819466-9.00001-8

important economic implications as well as social, safety, environmental, and sustainability implications, all of which are interdependent.

1.2 The role and constraints of process industries and their products

Process industries have a key role in transforming raw materials into finished products on a commercial scale. The processes involved typically require both physical and chemical changes, and in some cases require biochemical changes. The transformations are engineered within process plants. Most products of the process industries have well-defined specifications. Some products such as gasoline and beer, classified as *end products*, are used directly by the consumer but the majority, for example hydrogen, ethylene, and sulphuric acid, are classified as *intermediate products*, and are used in a vast network of downstream processes before the end product reaches the consumer. Some products such as catalysts or cleaning solutions are specialised in their applications, and their composition and properties reflect this. In all cases, product quality is of key importance, but the process of manufacture also assumes major importance because of the extensive raw materials and utilities consumed and the large capital and operating costs incurred. Utilities include electricity, fuels, water, cooling and heating media, and nitrogen for purging, and are heavily dependent on water and energy resources.

Process industries can be usefully classified based on the type of feedstock or end product involved, for example petroleum refining, mineral processing, chemical processing, fertilisers, food, and pharmaceuticals. National governments typically have a structured classification system for industries, enabling the collection of statistical data on related production, sales, employment, and investment.

Many industries provide service to process industries. One important example is the *process contracting industry*, which offers design, construction, commissioning, and project management services. Process contractors sometimes not only develop process technologies, but also enter into agreements with process licensors, often through research and development companies. Another important service industry is the *process equipment fabricating industry*, which designs and fabricates equipment items such as vessels, pumps, and heat exchangers as well as specialised equipment for particular applications. Further service industries include those that provide instrumentation and process control equipment and expertise. Services for industry devoted to technology research and development are available through both government funded and privately funded bodies. A range of consultants provide specialised services to process industries in diverse areas such as business management, technical safety, environmental protection, and information technology and computer software systems. All such service industries and consultancies need to have a fundamental appreciation of the cost structure and economic decision-making of the process industries. National governments have an important role in facilitating and regulating industries through policies on taxation, energy, safety, environment, and other key areas.

The variety of materials processed and produced by process industries is very extensive. Materials in solid, liquid, gaseous, and multiphase form are supplied in bulk quantities or packaged, and have diverse properties. Many are toxic, flammable, or explosive under certain conditions. Many, if not contained, present major safety hazards and also environmental hazards. Environmental impacts resulting from emissions include global warming leading to climate change, acidification, ozone depletion, eutrophication of waterways, and a range of toxicity impacts on humans and ecosystems. In producing marketable products, processes typically generate nonmarketable products or 'wastes', which must be reprocessed often by recycling, or treated and disposed of in an environmentally acceptable manner. Storage and transport of materials, often in large quantities and frequently under abnormal temperature and pressure conditions, are common practice. Safety and environmental hazards are thus key considerations in driving technologies, their implementation, and their costs.

Consumer products have a limited life after which they are traditionally disposed of as wastes. The major challenge here is to sort and recycle components of such products both to avoid environmental damage resulting from their disposal and to reuse valuable components, often involving further processing and refinement of the recycled materials. Such reuse serves to reduce both the depletion of extractable resources and the environmental impacts associated with their upstream extraction and processing. The concept of recycling materials at various stages of a product's life cycle, from extraction of raw materials to a product's end use, is often referred to as *circular economy*.

Process industries and their plants must be *economically sustainable*. Benefits derived from sales revenue must exceed operating costs and provide an adequate return on capital investment. Process plants are expensive to build in the first place, due to the materials and labour employed in construction, and the costs of planning, design, and project management. Plants are also expensive to operate due to the cost of raw materials, utilities, maintenance, insurance, and personnel employed, including management-related staff. Capital investment and operating costs must be competitive with those of other producers in the market place. A minimum scale of operation is usually necessary both to meet market demands and to achieve competitive processing costs.

1.3 Process industry projects

Fig. 1.1 depicts the anatomy of a process industry project. Following the identification of an investment opportunity, the evaluations of markets, available feedstocks, and appropriate technology will be undertaken. A key aspect in bringing these elements together successfully is the choice of a suitable site for manufacture. As the

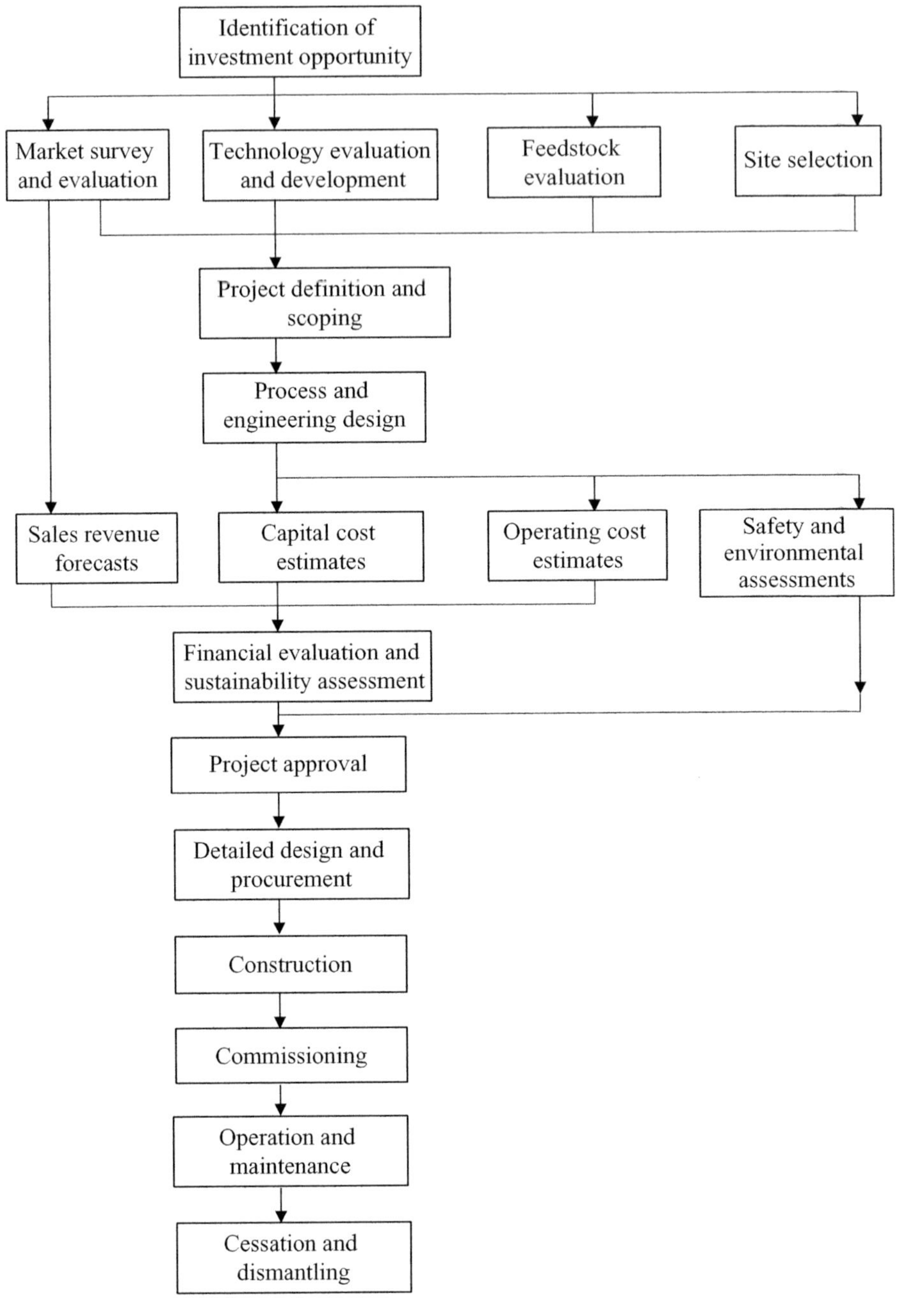

FIG. 1.1

The anatomy of a process industry project.

various elements of the project are refined and integrated, the project scope and characteristics can be defined. These include not only the markets, feedstocks, process technology, and site location, but also the production capacity, the extent of integration with other manufacturing plants, storage and transport of raw materials and products, the supply of utilities, and personnel requirements for design, construction, and operation. On the basis of this, more detailed market forecasts can be made, and the process and engineering design carried out to allow capital and operating cost estimates as well as safety and environmental appraisals to be made, and the overall sustainability of the project to be assessed.

During project evolution, interaction with government and community bodies must be initiated and the foundations laid for the acceptance of the project by the wider community. Satisfactory financial, safety, environmental, and sustainability performance will be key elements in gaining such approval but must be communicated to the wider community and particularly to those sections of the community that are directly affected. Project approval or rejection is normally the decision of the board of directors of a company who require a detailed, well-documented proposal, normally referred to as an *expenditure proposal*. A number of iterations of design and evaluation are often necessary before sanction is granted. Refinements to timing of investment, determination of plant production capacity, and other aspects of project definition are needed; these take time and effort by a range of professionals and thus involve costs.

The time span from identifying an investment opportunity to gaining sanction varies considerably; 2 years would not be unusual where the technology is established and market dynamics are reasonably stable, but longer periods are required if further research and development is necessary for the technology used or if further investigation and assessment of environmental and social impacts is needed. While Fig. 1.1 indicates the logical sequence of events, many iterations, and feedback loops are common within this framework. For example:

- safety assessments and environmental assessments can initiate revisions in site selection, technology selections, and design decisions influencing capital and operating costs;
- market evaluation may lead to a revised plant capacity decision, leading to changes in capital and operating costs as well as profitability;
- high operating costs whether in raw materials, utilities, or personnel employed may initiate design review leading to higher capital costs.

Major projects involve investment in a new process plant, sometimes on an existing industrial site (*brownfields*) and sometimes on an entirely new site (*greenfields*). However, there are many *minor projects* which involve modification to an existing operating plant. Such modifications may be driven, for example by:

- capacity expansion opportunities to take advantage of the market growth for the product;

- opportunities to reduce operating cost by improving efficiency in the use of raw materials or utilities, or by the addition of new equipment or improved process control;
- unforeseen changes in feedstock composition, product specification, or required operating conditions;
- availability of improved process technology;
- perception of safety or environmental risks which must be minimised or averted;
- tighter legislation regarding safety or the environment;
- opportunities to develop new links with other operating plants, whether existing or future, in order to achieve an improved industrial ecology.

In all of these cases, capital expenditure will be incurred and operating costs will change; in some cases, sales revenue increases. In all cases, economic evaluation will be required as part of the overall project evaluation, and a capital expenditure proposal will be required to enable authorisation of the proposed project. While minor projects inherently involve less capital expenditure than new projects, they must be carefully assessed across the spectrum of economic and sustainability criteria.

The time and effort spent on developing projects and the related process and engineering design as well as the related performance evaluations have significant implications for

- the personnel costs incurred;
- the extent of detail involved;
- the accuracy achieved in various performance evaluations, especially in
  - capital costs;
  - operating costs;
  - environmental impact;
  - safety.

Hence, there are a number of approaches to evaluation, which reflect the stage of the project's development, and the context of the evaluation. Thus, the cost of performing a study and achieving the necessary accuracy is somewhat less for a feasibility study than that required for a project at the sanction stage. This leads to a spectrum of approaches for capital and operating cost estimation which are applicable for different purposes and at various stages of project development.

1.4 Consumption and generation of funds

A major investment of capital funds is required for preliminary studies, project development, site acquisition, processing plant and related storage, buildings and utility generation facilities. At commissioning, additional capital is invested as start-up capital and as working capital for stocks of process materials and for the establishment of trading credit with suppliers and customers.

From the commencement of operation to the end of plant life, raw materials and utilities are consumed in the process, personnel are employed to manage and operate the plant and provide the necessary technical support, and material and personnel costs are incurred in the servicing and maintenance of the capital assets utilized. Collectively these comprise the production costs of the plant to which must be added the costs of research and development (or alternatively technology acquisition from an outside source), product distribution, customer support, and the corporate management of the business to make up the total operating costs. Revenue from product sales less the total operating costs generate a net income or profit, which hopefully provides an adequate return on the capital funds used.

There are some major calls on the ***net income*** generated from a project when it is in its operational phase. These include:

- meeting its *taxation commitments*; in this way the wealth generated by the project contributes to the wealth of the national economy through expenditure on education, health, defence, and so on;
- providing *dividends to shareholders* as a just and adequate return on their contributed funds; if funds have been borrowed from a lending source, then interest payments will be required;
- contributing to demands for *capital investment during the operating life of the plant*, when plant modifications are required;
- making *depreciation provisions* to enable eventual replacement of the plant at the end of its economic life; this is necessary to ensure long-term continuity in the same sphere of business activity;
- contributing to demands for *capital investment during the operating life of the plant*, when plant modifications are required.

Any surplus over and above these commitments can be set aside for longer term investment purposes, and for safeguarding against years of less successful operation or for unforeseen abnormal expenses, including those resulting from severe weather damage.

Capital investment in process industries makes very large demands on financial resources. Funds may come from within a company through retained earnings or shareholders' funds. Alternatively, they may be borrowed from lending authorities such as banks, insurance companies, or superannuation funds. In most cases, a mix of company funds and external borrowings will be used. From whatever source, funds have interest bearing capability and hence a cost. Thus, the profit generated must account for the cost of funds as interest, shareholder dividend commitments, or opportunity cost (i.e. the potential return from alternative investment opportunities). A generalised flow chart of money into and out of a process industry company is shown in Fig. 1.2. This flow chart is also applicable to a single project (as a segment of total company activity) contributing to the total wealth

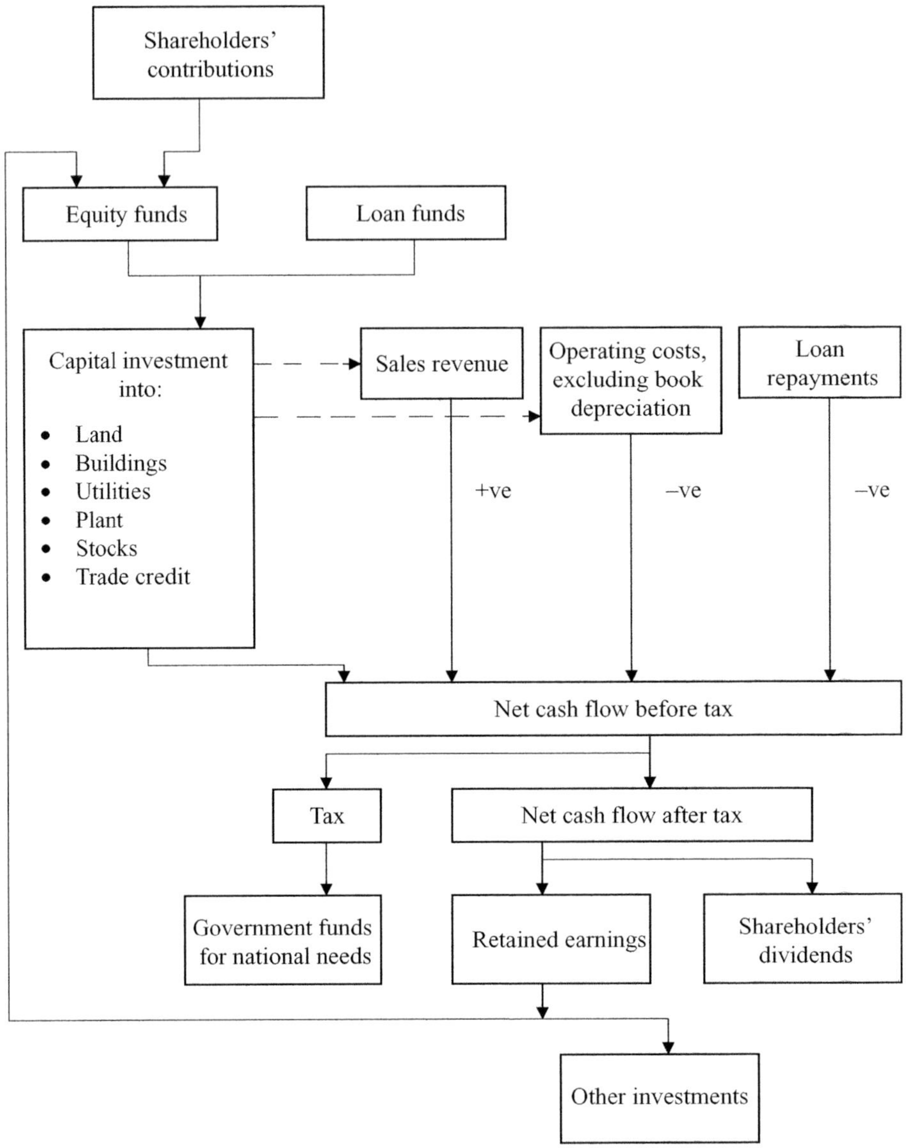

FIG. 1.2

Flow of money into and out of a process industry company.

and commitments of the company. Table 1.1 presents an example of a statement of wealth generated by a process industry company in a given year, showing separate contributions to employees, government, and finance providers, as well as money reinvested in the company's business.

Table 1.1 Statement of wealth created by Universal Chemicals P/L 2018.

	$ million	%
Wealth created:		
Sales revenue	1550	
Deduct cost of materials and services	1020	
Difference being wealth created	**530**	
Distribution of wealth created:		
• To employees in remuneration, retirement, and long service benefits including: Wages and salaries paid, company contributions to pension and provident funds, long service leave, and other employee benefits	**265**	50.0
• To government including Company income tax, payroll tax, and other taxes	**80**	15.1
• To providers of equity and loan capital including:		
– Interest paid less interest received	21	
– Shareholders' dividends	48	
– Minority interest in subsidiaries	9	
Subtotal	**78**	14.7
• Reinvestment in the business		
Depreciation	43	
Retained profit	64	
Subtotal	**107**	20.2
Total wealth distributed	**530**	100

1.5 Driving forces and influences for success

Driving forces for investment in new projects may originate from diverse sources such as:

- growth in market demand for established products;
- market demand for new or improved products;
- access to improved technology or the need to implement new technology;
- increase in prices of raw materials or utilities and the need to access new sources;
- strategic advantages, for example access to improved site location;
- demand for improved safety or environmental standards of processes or products;
- a strategic need to adopt improved sustainability in overall business and manufacturing structure.

Key determinants of economic success in the process industries are markets for products, availability of the necessary raw material and energy sources, competitive technology, a competitive cost structure, and appropriately skilled personnel. The process industries operate within a global economy. Products which can be safely transported can be made available at numerous locations in the world, providing potential competition to local manufacturers. Good access to markets and to basic raw materials and energy resources offer major economic advantages favouring certain locations for process plants. Particular benefits are achieved through economies of scale and through integration with other manufacturing facilities. Cheaper labour costs in many countries offer potential advantages in both reduced construction and operating costs for plants. Well integrated industrial facilities in mature economies can have major advantages over small-scale plants in developing countries. Access to large ports and transport infrastructure can have a major impact on transport costs, which can be a key component of raw material purchase prices and product distribution costs.

National and regional governments, besides deriving benefits from successful projects, can play a major role in stimulating and encouraging process industry investment. This is achieved, for example through policies on taxation, tariffs, royalties, investment incentives, and participation in infrastructure funding. Differences in government assistance or regulation can have a marked influence on variations in industry profitability from one country to another.

It is important to recognise the uncertainties associated with investment in process industry projects. The evaluation of markets, capital costs, and operating costs as well as the overall project profitability involve forecasting into the future over extended time periods. Dynamics of supply and demand patterns and related price dynamics, unforeseen changes in government policy or regulation, unforeseen technological changes, and actions by competitors, are all difficult to predict and can have strong influences on profitability. Unforeseen time delays to projects during design, construction, and commissioning, coupled with unforeseen difficulties in early plant operation can also exert significant influence.

Examples of potential failures in process industry plants affecting profitability include failure to achieve:

- designed production capacity;
- product quality requirements;
- required online time;
- targets for raw materials and utility consumptions and energy efficiency;
- required environmental performance in relation to emissions and wastes generated;
- required safety performance.

Some root causes of risks of technical performance failures include:

- where the project involves a pioneer plant employing new technology;
- the scaling up of a newly developed process;
- adapting previous design conditions to changed conditions, for example changed feed composition in a mineral processing plant;
- market pressure conflicting with adequate time limits for developing and proving the related process technology;

- operating complexity arising from interdependence of multiple plants, where unforeseen shutdowns or performance changes in member plants can impact the performance of a given plant.

1.6 The global context of the process industries

The process industries have become increasingly active internationally in recent decades with multinational companies operating in many parts of the world. Raw materials, intermediates, and finished products of the process industries are being traded internationally. Capital equipment items including construction materials, piping, vessels, complex machinery, and packaged plant can be sourced internationally.

An important variable in the context of international trade is the currency exchange rate between trading partners. The factors governing exchange rates are complex and outcomes are difficult to predict. For example, in some cases the currencies of countries have maintained their strength despite higher inflation rates than those of their trading partners. The influence of exchange rate on industries is likewise complex but it is useful to identify some of the influences.

For example, if we take a hypothetical trading relationship between Australia and the United States, then a fall of the Australian dollar relative to the US dollar:

- increases the value of Australian exports in Aust $ where export contracts are written in US $;
- increases the price in Aust $ of those raw materials, intermediates, and finished products imported from the United States to Australia at a given price in US $;
- increases the price of equipment items, instruments, and other components of process plant imported from the United States to Australia at a given price in US $.

The resulting increases in prices of raw materials or equipment imported from the United States for use in the Australian process industries increase both manufacturing costs and plant construction costs in Australia. Where a chemical product, which could be imported from the United States, is manufactured in Australia, the import parity price in Aust $ of the chemical product increases. Thus, an Australian manufacturer may increase the selling price of its product to an Australian purchaser, thereby increasing profitability. Historic exchange rates for selected international currencies is provided in Appendices.

International differences exist in relative labour costs, labour productivities, and access to raw materials and markets for products. International differences can also occur in occupational health and safety standards, in environmental standards and related policies such as carbon taxes, and in working conditions and capabilities of employees. Such differences have an impact not only on relative profitabilities, but also on supply security and quality of products.

1.7 Personnel organisation and interaction

Essential requirements in economic evaluation for a project can be identified as:

- market evaluation;
- capital cost estimation;
- operating cost estimation;
- cash flow projection over project life and profitability estimation;
- overall review of economics, cognisant of risks.

It is significant to note that these five key elements are almost always carried out by separate people with different skills and experience.

It is very important to recognise and understand how different people and skills are employed to evaluate and implement projects and operate the resulting facilities, how and by whom key decisions are made, and the procedures and time frameworks involved. An extensive range of disciplines and support skills are needed in design, plant construction, commissioning, operation, and maintenance of process plants. Chemical engineers are key players in these activities but other engineers are required in the design, construction, and management of projects. Boards of directors are responsible for allocating corporate funds to projects. Marketing personnel are needed for market evaluation. Financial, legal, and public relations expertise are necessary in numerous facets of the project evaluation and development. Various scientific disciplines are involved in technology development and management, where specialists in technology, safety, and environmental impact are needed.

In the *estimation of fixed capital costs*, professional capital cost estimators are often employed but build on expertise derived from the following specialists:

- mechanical engineers, having experience in the maintenance of equipment, machinery, and piping;
- civil engineers, having experience in designing foundations and structures;
- electrical engineers, having experience in the management of electricity supply, transformers, and electric motors;
- process control engineers, having experience in instrumentation and process control.

The experience of construction and project management personnel is also a key input into reliable capital cost estimation.

In the *estimation of operating costs* it is important to draw on the experience of technical and management personnel previously and currently involved in operations. In operation, a wide range of objectives are involved including:

- ensuring adequate availability to ensure reliable product supply to customers at the required rate;
- ensuring product specifications are met;
- ensuring safety of employees and the surrounding communities;
- meeting environmental requirements;

- ensuring that financial budgets are met;
- ensuring adequate plant maintenance.

Multiple personnel functions are involved including shift and day process operators, production managers, maintenance engineers and trades personnel, laboratory staff, development engineers, and sales personnel, as well as senior management and corporate specialists.

The various personnel involved in operations possess different levels of knowledge, experience, education, and skill. It is the task of management to draw together these skills harmoniously and productively. Different traditions develop in different industries, companies, and locations over time, and shape attitudes and management practices. Trade unions have traditionally had an important influence on wages, awards, and work attitudes. More recently, in many developed economies, there have been radical changes in workplace relations, and practices, with reduction in the number of employees in most businesses. There has also been a move towards multiskilling of labour and a departure from demarcation in areas of work responsibility. There has been a marked increase in female participation in most aspects of work, and this is likely to continue.

In the later stages of economic evaluation, it is important to consult with accountants and finance and tax specialists in assessing cash flows and profitability, and their dependence on taxation, regulation, and the financing (extent of equity and borrowed funds) of the project.

1.8 Ethics in project evaluation and management

Professional ethics are important in all aspects of project planning, implementation, and subsequent operation. The need for good ethical standards extends to ensuring:

- safety in construction, commissioning, operation, and decommissioning of plants;
- environmental standards in relation to resource consumption and emissions over the complete life of the project, including the ultimate dismantling of the plant and related site remediation;
- safety and environmental standards related to complete product life cycles, including product quality and responsible recycling or disposal of materials;
- social standards for those involved as employees over the extended life of the project including operations, as well as those in communities directly affected by the project;
- impartiality in weighing up market evaluation, technology assessment, capital, and operating cost inputs at the project evaluation stage;
- responsible stewardship of finances committed to the project for capital investment and operational expenses;
- transparency to shareholders, government, and the wider community regarding the essential aspects of the project and its financial, safety, environmental, social, and sustainability implications.

1.9 Outline of this book

From this chapter, the process industries can be seen to:

- be capital intensive;
- involve the use of valuable resources;
- demand a wide spectrum of skills in the evaluation and implementation of projects;
- have important implications for the environment, human safety, and sustainability;
- are subject to unforeseen changes, risks, and international competition.

Once operational, process plants and related businesses must be managed to secure the best use of the available assets. At both the investment and operational phases of a project, a sound understanding of the principles and methods of process economics is essential. The following chapters of this book are devoted to developing and applying these principles and methods encompassing:

- Chapter 2. Market Evaluation and Forecasting
- Chapter 3. Capital Cost Estimation
- Chapter 4. Operating Cost Estimation
- Chapter 5. Profitability Evaluation
- Chapter 6. Worked Examples in Process Economics
- Chapter 7. Process Technology Evolution and Adoption
- Chapter 8. Capital Investment Decisions
- Chapter 9. Industry Planning and Structure

In Chapter 2 the emphasis is on product classification, and the evaluation of market volume and selling price for feedstocks and products. Data sources for the production and price of both resources and products are explored.

Chapter 3 first examines the types of capital required for investment into projects. The distinction between inside and outside battery limits investments for fixed capital is explored. Various approaches to estimating fixed capital costs of process plants are then outlined. Working capital estimation is examined separately. Emphasis is placed on sourcing and critically reviewing cost data for equipment and plants.

Chapter 4 begins by distinguishing the boundaries and classification of operating costs. A simplified model for operating cost estimation is introduced. A more detailed examination is then made of the contributing elements to total operating costs. Effects of capacity utilisation and plant scale are explored. The sourcing and critical review of cost data is again tackled, in the context of operating costs; assumptions made in factored estimates are examined, for example in the light of related personnel requirements.

Chapter 5 explains how the evaluation of markets, capital, and operating costs collectively contribute to the cash flows over project life and the resulting project profitability. Cash flow and profitability estimation are developed, supported by worked examples. The effects of taxation, inflation, financing, and uncertainty are all examined. Approaches to uncertainty and risk are discussed.

Chapter 6 contains some problems and worked examples covering various aspects of cost estimation and profitability evaluation discussed in Chapters 2–5.

Chapters 2–5 provide a framework for evaluating process technologies and for understanding the driving forces for technology development. These driving forces are further examined in Chapter 7 in the context of technology evolution and adoption, exploring how market, technical, economic, and other dimensions can assist or deter technology development. Chapter 7 assists in understanding and quantifying the links between technology vintage and performance. The learning process in technology is explored, supported by quantitative measures. Learning is also explored in environmental, safety, and sustainability assessments through timelines.

Chapter 8 deals with several categories of capital investment decision driven by different controlling influences involving different perspectives in their evaluation. Capacity expansion through both new and existing plant is examined in some detail. The links between the evaluation of economics and that of safety, environmental, and sustainability performance are explored, supported by case examples.

Chapter 9 considers some implications of process economics for industry planning, structure, and performance. Aspects of industry classification, characterisation, integration, location, ownership, and performance are explored. Some organisational relationships of industry with some key community bodies are discussed.

The following objectives are interwoven throughout the book:

- provide and integrate all the tools required for the economic evaluation of a new project or an existing operation;
- reflect the breadth and diversity of personnel, skills, and approaches necessary for economic evaluation;
- identify the relationships between technology and engineering with costs incurred or revenues generated;
- place cost estimation and economic evaluation within a range of contexts of project evaluation and plant management, recognising that the context has major implications for the extent of detail, time required, and achievable accuracy for the evaluation;
- critically review economic data and assessments;
- explore the links between economic evaluation and sustainability, environmental, and safety assessment;
- encourage further reading of the literature and thoughtful consideration of the factors underpinning the practice of economic evaluation and decision-making.

CHAPTER

Market evaluation and forecasting

2

Everyone lives by selling something.
Robert Louis Stevenson in Across the Plains (1892)

Chapter outline

Process Industry Economics. https://doi.org/10.1016/B978-0-12-819466-9.00002-X

2.1 The importance of markets

Market evaluation is a key component of economic evaluation in the process industries. The market place has a critical role in determining the success of particular industries and of individual projects within those industries. Markets impact on both *purchasers* in influencing prices of raw materials, utilities, and labour, and *sellers* in determining the revenues from sale of resources or manufactured products, or from services rendered. Sales revenue for products depends on *sales volume* and *selling price*. Since major projects typically involve significant lead times from planning to authorisation, through to design, construction and commissioning, followed by extended operating lives, market evaluation involves not only assessment of the present situation, but also forecasting into the future. Markets are important in a number of contexts.

2.1.1 The financial context

Sales revenue, the product of sales volume and selling price, is invariably a key driving force for business success. Sensitivity studies within profitability evaluations for process industry projects have consistently identified product selling price and market volume as the most influential parameters affecting the calculated measure of profitability [1, 2].

2.1.2 The chemical engineering context

Forecasts of sales volume for a new project have a direct bearing on the choice of *production capacity* for the associated plant. Inadequate capacity has major ramifications in terms of foregone sales opportunity. Excess capacity is reflected in surplus capital investment, and restricted capacity utilisation in operation resulting in increased fixed operating costs per tonne of product.

Specification of *product quality* (and *packaging*, if relevant) influences process and equipment design, and hence capital and operating costs. Product quality is vital to product performance, its competitive position in the market place, and sales revenue.

Thus the chemical engineer is dependent on good quality market information as an input into design decisions. At the same time, the chemical engineer can make a worthwhile contribution to market appraisal and product design through technical knowledge of processes and the associated industry structure. Since product selling prices must cover operating costs to ensure profitability, engineering cost estimates also play an important role in assisting market negotiations on product pricing.

2.1.3 The environmental, safety, and sustainability context

As well as being evaluated in the financial context, products are evaluated for their environmental impact assessed over their entire life cycle, encompassing:

- extraction and purification of resources required for manufacture
- the various stages of processing in product manufacture
- product use and disposal
- potential for materials recycling at various stages of the product life cycle

This evaluation includes assessment of resource depletion and impacts derived from emissions and wastes generated, as well as the potential for materials recycling. The inherent safety attributes of the various steps of the life cycle chain, including materials storage and transport, are also important. Product vendors have a specific responsibility in ensuring the customer is aware of any hazards and of necessary safe practices in the transport, storage, and use of a product.

2.2 Uncertainties in market forecasting

Whilst market forecasting is important to the success of process industry business, it is also fraught with uncertainty. Uncertainty is derived from many sources including

- business cycle fluctuations which can be caused by a wide variety of influences
- changes in process technology or product development
- changes in industry structure
- changes in international participation in manufacturing
- changes in the balance of supply and demand
- changes to international trade arrangements
- changes in environmental drivers, including government regulation, for example on global warming impacts, materials recycling

Macroeconomic growth and competition are two important business influences. Macroeconomic growth in a particular country can influence its consumption patterns as well as its capacity for investment in process plants and in research and development.

The Australian government report 'Resources and Energy Quarterly' [3], for example, provides an insight into the macroeconomic effects which influence potential sales volumes of exported mineral and fuel commodities from Australia. These influences include trends in

- gross domestic product trends in purchasing countries
- national policies on taxation and their effects on investment and employment
- tariffs and their effects on international trade
- inflation rates and currency exchange rates in different countries

Competition can occur both where multiple traders or manufacturers sell the same product (e.g. methanol and sulphuric acid) and where manufacturers sell competing products having the same function (e.g. glass bottles compete with aluminium cans in the drink container market, soda ash competes with caustic soda for some alkali markets in chemical processing). Competition can occur within a country (e.g. within

Australia), a region (e.g. between different countries in Europe), or internationally (e.g. between China and the United States). The manufacturing and marketing strategies of a competitor, which are always subject to uncertainty, can have major impacts on the market position of a product supplier. Competitors may be either established participators or newcomers in a market. There is always an element of unpredictability about competitor behaviour; apart from inherently human characteristics and political influences, one cannot realistically know all the constraints, opportunities, and aspirations of a competitor.

Despite the many uncertainties in market forecasting however, a thorough and detailed market study is an essential component of project definition and evaluation for a new business venture, and a key element of managing current business operations.

2.3 Product classification

The process industries can be broadly viewed as transforming fuel, mineral, and agricultural commodities through a network of physical and chemical processing into finished products. Other manufacturing industries then convert the products of the process industries, typically by fabrication and assembly, into engineering or consumer products. Examples include

- extraction of bauxite, conversion of bauxite to alumina, conversion of alumina to aluminium, and fabrication of various aluminium products
- extraction and purification of natural gas, conversion of methane to hydrogen, conversion of hydrogen to ammonia, and conversion of ammonia to a range of fertiliser products
- extraction of ethane from natural gas, its conversion into ethylene, conversion of ethylene into polyethylene, and fabrication of a number of polyethylene products
- extraction and refining of crude oil into a range of fuel products such as gasoline and diesel, and also into feedstocks such as naphtha and sulphur for further processing into downstream products

A number of terminologies are used to distinguish various classes of products within the process industries. An important distinction, for example, is made between *consumer goods* and *producer goods*.

A consumer good is a product which requires no further processing prior to use by the ultimate consumer. Toothpaste and beer are examples of consumer goods. In consumer market research, the product end user is readily identifiable. Accurate sampling of the population is possible from well-documented statistics on the size and distribution of the population. From interviewing a statistical sample of consumers, a reliable estimate of market preferences and requirements can be made.

Producer goods are at least one stage (and normally much further) removed in the processing sequence *upstream* from the consumer. A producer good such as ethylene diffuses through a network of processing and distribution activities before eventually

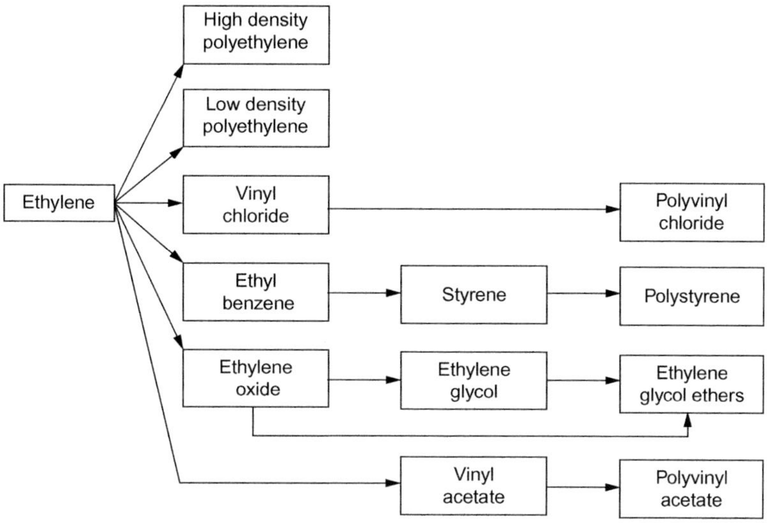

FIG. 2.1

Part of the diffusion of ethylene through the process industries.

reaching the consumer in products as diverse as plastic wrap or antifreeze. The diffusion network is characteristically complex and extensive. Fig. 2.1 provides a simplified view of the diffusion of ethylene to intermediate and downstream products.

A further important distinction is made between *differentiated* and *undifferentiated* products. An undifferentiated product has a specific chemical composition and particular physical specifications regardless of its producer. A differentiated product has a distinguishing feature compared with similar products made by a competitor, and is usually sold on the basis of its performance. Product differentiation may be achieved by chemical and physical property differences, and also by more subtle differences such as method of packaging, provision of an additive, or even elements of technical service.

By adopting a further classification by production volume, a simple matrix can be postulated for the chemical industry distinguishing 'true' commodities, 'pseudo' commodities, fine chemicals, and specialty chemicals (see Table 2.1). This approach to product classification and its use in market forecasting approaches has been discussed by Kline [4]. Table 2.2 provides more detailed characteristics of these product classes.

Table 2.1 Classification of chemical industry products.

Annual production volume	Undifferentiated	Differentiated
High	Commodity	Pseudo-commodity
Low	Fine chemical	Specialty chemical

Table 2.2 Characteristics of product classes.

Commodity
Product traded universally to same specification
Synthesised in high volume, frequently from captive raw materials
Widely used by many customers, often with the greater part of sales volume concentrated in a small number of purchasers
Fine chemical
Undifferentiated
Synthesised in low volume
Often sold to small number of customers, each buying in moderate to low volumes
Usually produced for one or more end uses to established standards (e.g. pharmaceutical and food codes)
Pseudo-commodity
Synthesised in large volume, often from captive raw materials
Bulk of sales often concentrated in few large customers
Differentiated
Produced to performance rather than composition specification
Specialty chemical
Differentiated
Synthesised or formulated in low volume
Designed to solve specific customer problems
Often distributed to relatively large numbers of customers, each buying in relatively small volume

The term *commodity* is used widely to indicate a valuable product which is traded internationally. Examples include

- **fuel resources** such as coal, natural gas, and crude oil
- **mineral resources** such as bauxite, iron ore, and phosphate rock
- **metal products** such as copper, zinc, and steel
- **agricultural products** such as barley, wheat, and sugar

The term *commodity* is also used to describe a wide range of chemical products traded internationally. Examples of commodity chemicals include ethylene, ammonia, and sulphuric acid. The world demand for commodities is very large, encouraging large-scale manufacture, frequently from captive raw materials. The success of a commodity business depends on an ability to supply a consistent quality product at a competitive price and in agreed quantities, within an agreed time framework.

As markets have developed and matured, they have tended towards an increased degree of differentiation. Thus polymers like polyethylene and polyvinyl chloride, which may once have been regarded as commodities, have become increasingly differentiated and can now be regarded as pseudo-commodities. Differentiation has been achieved by modifying process conditions to provide a diverse range of specifications which meet the various needs of customer applications.

Examples of fine chemicals include products like aspirin, tartaric acid, and citric acid. Fine chemicals are made in lower volumes and to tighter product specifications than commodities. Many are food additives and pharmaceutical intermediates.

Examples of specialty chemicals include catalysts, water treatment aids, mineral processing aids, oilfield chemicals, pulp and paper chemicals, and pharmaceutical end products. Specialty chemicals are developed to meet specific customer needs, and are described by their functions rather than product names. Being tailored to specific needs, they are highly differentiated and tend to be produced in relatively small volumes. They also incur a high degree of technical service in use, and research and development in ongoing product improvement. They are often distributed to relatively large numbers of customers, each buying in relatively small volumes.

Other ways in which products may be classified include the relationship between price and volume for various scopes of product classes. A more detailed discussion of product classification is given by Wei et al. [5] and Couper et al. [6].

Product classification is useful in assisting an understanding of the distinctions between the business characteristics related to different types of products, not only as they affect market forecasting and strategy but also as they affect strategies for manufacture, and for research and development. Taking commodities and specialty chemicals as extreme examples of the product classification spectrum, these distinctions are outlined in Table 2.3. A number of process design texts, for example Smith [7] and Seider et al. [8], give insight into the different manufacturing strategies adopted for high-volume and low-volume products. Seider et al. also give insights

Table 2.3 Key business characteristics of commodities and specialty chemicals.

Business characteristic	Commodities	Specialty chemicals
Research and development	Process technology	Application and field testing
Manufacturing	Large capacity, dedicated plants	Small, flexible multiproduct plants
Marketing	Large market volumes	Small market volumes
	Few large customers	Many customers
	Little technical service	High degree of technical service
	Little test marketing	High degree of test marketing
	Price sensitive to business cycle influences	Price robust to business cycle influences
	Extensive market research geared to volume and price forecasting	Market research geared to individual customer needs

into strategies for '*configured' consumer products* sold directly to consumers using examples as diverse as halogen light bulbs, haemodialysis devices, soap bars, ice cream, and cheese substitutes.

2.4 Estimation of market volume

The estimation of market volume can be made by several methods, involving different perspectives, degrees of detail, and incurred costs. Since estimating approaches are necessarily imprecise, it is all the more valuable to make estimates from a number of perspectives.

2.4.1 End use analysis

This method identifies end uses (existing and potential) and projects them, after studying the factors influencing demand in each sector of the market. Let us take the example of low-density polyethylene (LDPE), and assume that the end uses comprise:

- 34% film and sheet, comprising 26% in packaging and 8% in nonpackaging (such as construction, household, and agriculture)
- 26% injection moulding
- 40% other applications (e.g. co-polymerisation with esters, extrusion coatings for paper and other substrates, wire and cable insulation, chloro-sulphonation to Hypalon)

The demand in each sector of the market (such as film/sheet packaging) is first determined in order to establish the extent of penetration by LDPE, and to then estimate the future penetration rate by LDPE in the light of expected price, production cost, and technology trends. This is done for the whole market, accounting for the effects of new materials which may compete, technology changes in user industries, and potential for new uses of LDPE.

If we were to make such a study for ethylene, the end uses would be far more complex and numerous, with the need to consider all derivatives including

- polyethylene (low density, linear low density, and high density)
- ethylene glycol and other ethylene oxide derivatives
- vinyl chloride and PVC
- styrene and polystyrene
- vinyl acetate

These derivatives diffuse through the industrial processing network to end up as fabricated plastics, antifreeze, apparel fibres, solvents, and other products.

In end use analysis, it is desirable to conduct field surveys to discuss trends in consumption patterns with manufacturers (including historic data and future estimates) as well as possible changes in materials and production techniques. Such field surveys are time consuming and costly, but achieve a detailed assessment of the market. Errors in various sectors are unlikely to be cumulative (some will involve overestimates, some underestimates); hence the technique is potentially reliable.

In determining the potential end uses of an intermediate chemical, it is necessary to have a good understanding of the commercial process routes using that intermediate as a feedstock, as well as the relevant commercial routes for downstream processing. An understanding of industry structure within strategically important regions is also necessary. Useful guides to the potential network of chemicals via commercial processing routes are given by

- Siggurdson and Rudd [9], Seddon [10], and Wells [11] in relation to petrochemicals
- Kirk Othmer [12] and Ullmann's [13] encyclopaedias in relation to chemicals generally
- Gary, Handwerk, and Kaiser [14] in relation to petroleum refining
- Australasian Institute of Mining and Metallurgy [15] in relation to minerals processing

Useful contributions to understanding the structure of the chemical process industries are provided by Wei et al. [5] and Couper et al. [6].

2.4.2 Statistical analysis and projection

Statistical techniques can be applied to historical data for a product in the form of annual consumption or annual per capita consumption over time to establish a trend. The trend is then extrapolated into the future. The difficulty is in identifying the point already reached on the life cycle curve, and in making the extrapolation.

The life cycle curve, normally represented as a sigmoid (or S) curve and shown diagrammatically in Fig. 2.2, is helpful in representing the development of product growth.

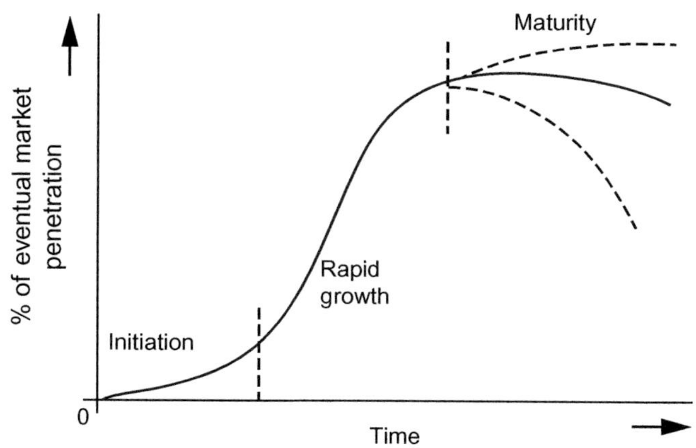

FIG. 2.2

Generalised life cycle curve for a chemical product.

Three growth phases can be postulated:

Initiation phase: involving introduction of the product to the market. Technical problems relating to the product and its manufacture must be overcome. The period is marked by intensive market development activity and relatively slow growth.
Rapid growth phase: involving growing market acceptance. As market growth occurs, manufacturing plants become larger, economies of scale and technological improvements are achieved leading to production cost reduction; cost reduction encourages market growth and there is a cyclic process sustaining market growth until some limiting factor retards the growth process.
Maturity: growth slows and may cease showing a levelling of demand and may even decline, especially if newer competing products penetrate the market.

The product life cycle curves are often represented as logistic or Gompertz curves. The logistic curve may be represented by

$$D = \frac{D_u}{1 + A\exp(-Bt)} \tag{2.1}$$

whilst the Gompertz curve is represented by

$$\ln D = \ln D_u - A\exp(-Bt) \tag{2.2}$$

where

D = demand at time t
D_u = ultimate demand
t = time in years
A, B = parameters

Linear and exponential functions have also been found useful in representing product growth. Historically, many products have experienced growth which could be modelled by an exponential growth relationship, that is

$$D = D_i \exp(zt) \tag{2.3}$$

where

D = demand in year t
D_i = initial demand
z = exponential growth rate

Thus industrialists have expressed market growth in terms of 'growth rates' implying exponential growth. Production, distinguishable from consumption because of import and export trade, is nevertheless a useful indicator of product demand.

Influences which constrain growth include excessive production costs or selling prices for a product, technological limitations to further product or process development, scale limits in equipment and entire plants, and health, safety, or environmental factors relating to specific products or processes. Price increases in crude oil in 1973 and 1979 led to increased costs (and hence prices) of products based on hydrocarbon feedstocks or on energy intensive processes. Increased product prices did much to

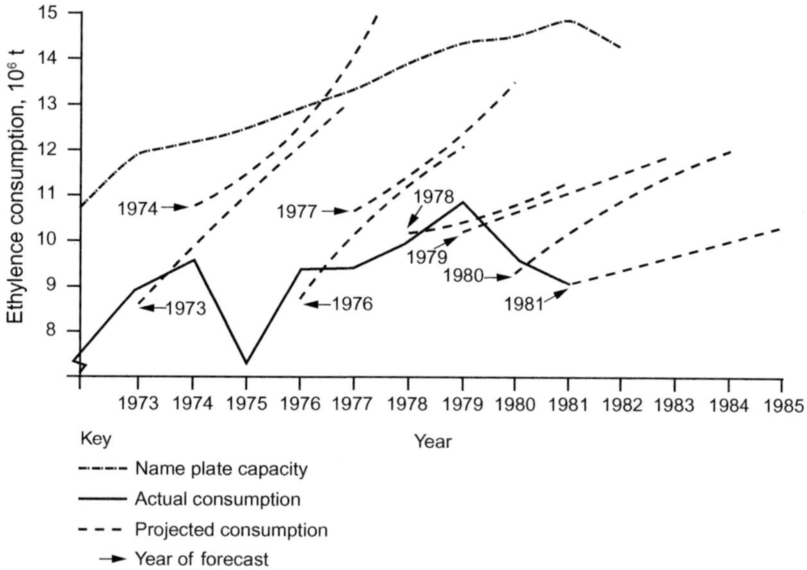

FIG. 2.3

Projected and actual consumption of ethylene in the EEC.

From G.S.G. Beveridge, The Institution of Chemical Engineers Presidential Address, 1984-85.

dampen demand, with greatly reduced market growth rates from 1973 to the early 1980s. Fig. 2.3, which shows the comparison between actual and projected ethylene consumption in the European Economic Community, may appear as a damning indictment of market forecasting methodology. However in the context of the aftermath of sustained and high exponential growth over two to three decades, the projections are at least understandable. Interestingly these overestimates of growth contrast with underestimates of ethylene growth rates in the 1950s and up to the mid-1960s [16].

Another important factor influencing the slowing of growth for many base chemicals has been the growing awareness of the environmental impacts of emissions and wastes generated and of resource depletion, and related pressures to improve the efficiency of feedstock and energy resource utilisation. In extreme cases, environmental or health factors can cause the complete demise of a product's market position, as occurred with chlorinated fluorocarbons arising from global commitment to addressing ozone layer depletion in the 1990s. Increasing attention has more recently focussed on global warming and on the need for mitigation within the industry. The IPCC Climate change 2014 report on industry [17] identified steam cracking of hydrocarbons for light olefins production such as ethylene and propylene as the most energy consuming process in the chemical industry. Other major contributors cited were production of ammonia, nitric acid, adipic acid, and caprolactam. Impacts result from emissions derived from energy consumption and by-product venting. Environmental concerns are also increasing focus on the potential use of biomass materials for fuels and chemical feedstocks.

2.4.3 Correlations with economic indicators

Since commodities diffuse so extensively through the industrial economy, it has often been possible in mature large economies like those of the United States and United Kingdom to relate growth in the volume of demand for a particular product to basic economic aggregates such as:

- gross domestic product
- capital investment
- industrial production

Published indices or forecasts for these economic aggregates can then be used as indicators of a product's potential market demand. Other economic indicators for countries of importance for markets of basic fuel and mineral resource commodities include a country's level of debt, prevailing interest rates, and extent of employment.

It is also possible to make similar links between demand for particular products with business indicators for a relevant sector of the economy. Demand for PVC for example can be linked with the health of the domestic and industrial building industry.

2.4.4 Comparison of per capita consumptions

By selecting countries at a more advanced level of industrialisation, a time lag can often be established between relative per capita consumption in the more and less industrially developed countries. Current per capita consumption in countries from mature economies may also be useful as a guide to the potential per capita consumption in a developing country, for example. One needs to make allowance for factors such as climate and lifestyles in the countries concerned, as well as relative patterns of industrial development and relative industry structures.

2.5 Selling price estimation and forecasting

Selling prices for products can move erratically and apparently unpredictably at times, making price forecasting for products a most difficult task. Despite this, some concepts are useful.

One concept, discussed by Rudd and Watson [18] in the context of accommodating to future developments, is that of a product's selling price comprising a ***floor price*** and a ***margin***. The floor price is the minimum selling price required to provide a minimum return on capital investment and to cover the production cost based on best available technology and plant scale, as well as competitive unit costs. The product can be sold at a margin above floor price depending on the level of competition. For a new product, the margin can be considerable and needs to compensate for research and development costs and the risks incurred in bringing the new product to the market. As competitors emerge, attracted by prospects of high returns, the margin decays. The relationship can be expressed as

$$P = F\exp(-K_{F}t) + M\exp(-K_{M}t) \tag{2.4}$$

where

P = product selling price in real terms (\$/tonne)
F = floor price in real terms (\$/tonne)
M = margin over the floor price in real terms (\$/tonne)
t = time in years
k_F k_M = decay parameter

Both the floor price and the margin can be expected to decay, in real terms, with time. The floor price decays under the influence of improved technology, increased scale of production, and increased skill in production and management. The margin decays under the influences of increased competition and hence reduced market share. Typical rates of decay suggested by Rudd and Watson were 1%–4% per annum for floor price, and 10%–20% per annum for the margin [18]. Rates of change in production cost due to technology improvements and operational learning are discussed in Chapter 7 in the context of technology evolution.

Some products from small-scale plants may be sold at levels above the floor price when protected by remoteness and the transportation costs of competing products from their point of manufacture. Other forms of protection include tariffs and other trade restrictions imposed by national governments, as well as currency exchange rates between importing and exporting countries.

For a product manufactured in a country where imports are permitted, the maximum selling price achievable is fixed by the import parity. Import parity is the minimum price for which the product can be landed when imported from another country. The factors influencing an import parity price are

- source of the import which may be Europe, the United States, the Middle East, China, or elsewhere
- any applicable tariffs, which function as a percentage of FOB price (FOB = freight on board, which is the cost of the product transported to the ship's cargo in the exporting country)
- freight (which is a function of distance, tonnage, packaging and shipping frequency) as well as insurance and landing charges
- currency exchange rate

As an example, the import parity price per tonne of product of a polymer imported into Australia from Europe in the context of a hypothetical tariff could be

FOB price	1200 euros per tonne
Duty (say at 10% FOB price)	120
Freight (say)	160
Landing (say)	20
Total	1500 euros per tonne

When converted to Australian dollars, this becomes \$2400/tonne for an exchange rate of 1.6 Aust\$ to the euro.

Currency exchange rates exert a strong influence in determining import parity prices. The uncertainty regarding future trends in exchange rates, as well as sources of import competition, makes the forecasting of import parity difficult. Operating costs of manufacturers in a given country may move in different trends to those of overseas competitors in relation to feedstock, energy, labour, and freight costs. Differences in capital investment and fixed operating costs may also differ. Despite the complexities, some competitive analysis of world manufacturers is essential to price forecasting.

There are some cases where prices are effectively immune from import competition. These occur when products are either too toxic (e.g., liquid chlorine) or where transport costs are too high relative to product value to permit transportation over large international distances. It is important to remember however that the derivatives of such products may well be subject to import competition, and that weaknesses in the cost structure of upstream processes will play a part in the competitiveness of the end product.

Whilst import parity pricing effectively sets one upper boundary for a manufacturer's selling price, additional competition may exist from other manufacturers within the same country, driving selling prices below import parity levels.

Prices are influenced worldwide by the balance of supply and demand factors. In periods of excess of production capacity over demand, whether in extraction facilities for resources (minerals, oil, natural gas) or in manufacturing plants, suppliers are forced to lower their selling prices to achieve sales. Considering Eq. (2.4), suppliers can lower their selling price to the floor price with zero margin, and may be prepared to continue supplying for short periods at even lower price levels with the expectation of a more favourable trading position when demand strengthens. Similarly, in periods of supply capacity shortage, manufacturers can raise their prices.

2.5.1 Price elasticity

The relationship between selling price and demand for a product is thus an important aspect of product price behaviour. The relationship is termed *price elasticity*. If the demand for a product is independent of price, the demand is inelastic; if demand is highly dependent on price, the demand is elastic. This property can be quantified by the price elasticity of demand E where

$$E = -\left(\frac{\mathrm{d}V}{V}\right)/\left(\frac{\mathrm{d}P}{P}\right) \tag{2.5}$$

where

V = volume of demand over a fixed period of time
P = selling price of product
dV, dP indicate differential changes in volume or selling price and the value of E is always positive

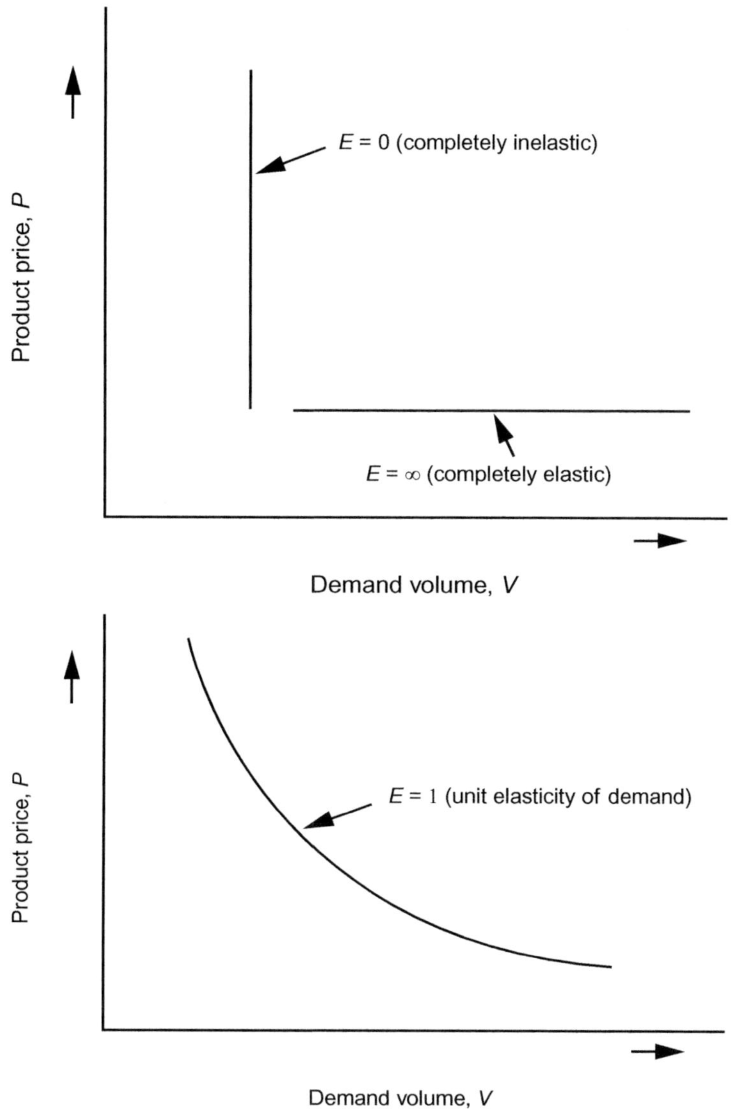

FIG. 2.4

Illustration of the price elasticity of demand property.

The property of price elasticity is illustrated in Fig. 2.4. Elasticity is not always easy to determine or forecast, however.

As Wei, Russell, and Swartzlander [5] point out, an important influence on price elasticity of demand is the availability of a substitute for the product. If we consider glass bottles for beer containers for example, a sudden increase in price might be followed by adoption of aluminium cans as substitutes so that the price elasticity is large. If we consider petrol derived from crude oil, alternative fuels such as diesel might be attractive for alternative vehicles but the market is constrained by lack of flexibility in the ability of a specific car to adopt an alternative fuel. Fuel ethanol might be an attractive alternative, but depends on the availability of feedstock and cost of production. Longer-term forecasts of escalating crude oil prices or reduced refining capacity might encourage investment in electric or hybrid vehicles, but such change requires vehicle replacement with associated cost. More fuel-efficient cars and increased adoption of public transport may take some time to eventuate.

Following the crude oil price increases in the 1970s by the OPEC countries of the Middle East, much world investment was diverted into establishing alternative sources of hydrocarbons and fuels. Investment was also directed into making fuel combustion engines and processes in refineries and petrochemical plants more energy efficient. Many fuel oil-fired boilers were replaced with coal-fired boilers. Thus the overall demand for crude oil, and for crude oil supplied by OPEC countries, was dampened, making it impossible to sustain high crude oil prices. After a steep increase in the real crude oil prices in the 1970s, there was a reduction in real prices in the 1980s, levelling out in the 1990s.

There is thus a distinction between short-term elasticities and long-term elasticities which depend on readjustment in industrial planning, technology use, and operating strategies as well as on personal lifestyles.

A distinction is made by economists between elasticity of demand and elasticity of supply, distinguishing between the effects of changes in prices on quantities demanded or supplied. Of interest in the process industries is the effects on price of oversupply or undersupply of a particular raw material or product.

An oversupply of a product often results from excessive investment in new plant capacity based on an expectation of demand growth which fails to eventuate. Investment decisions characteristically have a lead time of 2–3 years from sanction to plant commissioning; a further lead time of several years may occur before market demand matches plant capacity. These protracted lead times contribute to forecasting errors. The early 1980s saw excessive manufacturing capacity in petrochemicals worldwide resulting from over-optimistic forecasts. This led to strong competition between suppliers for markets, and a consequent reduction in product selling prices. The adjustment phase which followed required time for re-establishment of market growth, and saw the closure of less competitive plants prior to improved capacity utilisation enabling restoration of higher selling prices.

2.5.2 Price histories for raw materials and products

Selling price histories for fuel and mineral commodities show considerable volatility over time. This volatility reflects

- variation in supply in relation to demand
- decline in production capability of current supply facilities
- variation in the numbers of new supply sources becoming available

- variation in demand
- variations in the degree of activity in the broader economy

A new mine or oil or gas field requires significant lead time in terms of exploration, assessment of quality and extent of reserves, assessment of environmental issues and economics, gaining government approval, obtaining finance, as well as the project time to implement the planning, design, construction and commissioning of the extraction, purification, and shipping facility.

Selling prices for products can also show a degree of volatility due to

- variation in demand by downstream manufacturers
- variation in production capacity caused by temporary or permanent closure of existing facilities, or new capacity coming on stream
- extent of competition between suppliers arising for example from new entrants to the market with operating cost advantages

2.5.3 Assessment of selling price data

Price data for selected fuel commodities, raw materials, and products are published by government organisations, technical and business journals, and business consultants. It is important to consider some important aspects when interpreting these data. These include the following:

- Prices may be reported on a spot or contract basis:
    - a contract basis may involve a specific supplier over a specific supply period, which may extend for many years
    - a spot price, negotiated for immediate or early delivery, is typically for smaller sales volumes over short-term periods
- The price of many products will depend on how the product is supplied, for example by pipeline, by shipment with bulk storage, and whether the product is packaged
- The price will depend on the degree to which freight costs are included, for example:
    - an ex-works price (freight excluded)
    - an FOB (free on board) price reflecting the price on board ship in the country of origin, where the importer pays subsequent transport costs to the destination
    - a CIF (cost-insurance-freight) price which includes the costs of all freight and distribution, including insurance, incurred in supplying the customer
- where a product is packaged, the price will depend on the size of the container: this applies for gases such as hydrogen and liquefied gases such as liquid oxygen, liquid nitrogen, or liquefied petroleum gas
- the price will depend on the country of origin, and within individual countries may also depend on the specific supply location, reflecting costs of extraction and/or manufacture

For the case of natural gas sales, different types of contracts for both pipeline gas and liquefied natural gas are discussed in some depth by Chandra [19].

2.6 Resource production, consumption, and reserves

Fuel commodities are important both as sources of utilities and feedstocks for the process industries. Natural gas has clean burning characteristics and a high fuel combustion temperature, and is used both for process heating and steam and electricity generation. Natural gas is also an important feedstock for many chemicals such as hydrogen, methanol, and ammonia, and its price impacts on the prices of these chemicals and their derivatives.

Crude oil is refined to produce a range of fuels (e.g. aviation fuel, gasoline, diesel, fuel oil) and also hydrocarbon feedstocks such as naphtha for chemical processing. Coal has had a key role both in power generation and in specialist applications such as providing coke for iron and steel production.

Mineral commodities are important as a source of metals. Bauxite is first converted to alumina by treatment with caustic soda followed by calcining; alumina is subsequently smelted using electrochemistry to produce aluminium. Many metals such as nickel, copper, lead, and zinc occur naturally as sulphides; during smelting of the sulphide ores, sulphur dioxide is released and is normally captured to make sulphuric acid, which has an important role in many chemical and mineral processes. Thus there is an interdependence of resources and their extraction with the wider process industries.

Materials recycling at various stages of product life cycles has an influence on consumption of raw materials from which those products are derived. The existence of a material processing life cycle also often provides opportunity for used products to be recycled back into the manufacturing chain; an example is used glass or 'cullet' which can lower furnace melting temperatures and hence fuel consumption when recycled to supplement traditional feed materials such as sand, limestone, and soda ash.

2.7 Data sources for production and prices of fuels, raw materials, and products

Table 2.4 summarises some data for consumption, production, and reserves of key fossil fuel commodities as of 2017 for the world and some dominant nations in their role as producers or consumers. Some indicative prices are also shown. Data have been drawn from the BP Statistical Review of World Energy [20] which provides detailed reports of consumption, production, and reserves for coal, crude oil, and natural gas for most countries, as well as data on different energy sources, carbon dioxide emissions, and world prices.

Considering the case of natural gas, aspects of importance include those derived from

- Changes in consumption rates over time, for example world consumption has increased at a rate of 2.3% per year, but at higher rates in some individual

Table 2.4 Some data for world consumption and production of fuel commodities [20].

Fuel	Production 2017	Consumption 2017	Consumption growth rate	Price 2017	
Natural gas	**Billion cubic metres**	**Billion cubic metres**	**%/year (2006–16)**	**US$/GJ**	**Notes**
World	3680	3670	2.3		
Russia	636	425	0.0		
United States	735	740	2.5	3.0	Henry Hub
United Kingdom	42	79	−1.5	5.8	Heren NBP index
Australia	114	42	4.9		
China	149	240	13.7		
Qatar	176	47	8.3		
Japan		117	2.9	8.1	Henry Hub + liquefy $3 + ship $2.1
India	29	54	3.5		
Europe	242	532	−0.9	5.6	
Fuel	**Production 2017**	**Consumption 2017**	**Consumption growth rate**	**Price 2017**	
Crude oil	**Million tonnes**	**Million tonnes oil equivalent**	**%/year (2006–16)**	**Spot US$/bbl**	**Notes**
World	4387	4622	1.1		
Russia	554	153	1.3		
United States	571	913	−0.7	51	West Texas
Canada	236	109	0.4		
Europe	163	731	−1.3		
Australia	15	52	1.2		
India	40	222	5.2		
China	195	608	5		
Middle East	1481	420	2.7	53	Dubai

Continued

Table 2.4 Some data for world consumption and production of fuel commodities [20]—*cont'd*

Fuel	Production 2017	Consumption 2017	Consumption growth rate	Price 2017	
Coal	**Million tonnes oil equivalent**	**Million tonnes oil equivalent**	**%/year (2006–16)**	**Spot US$/tonne**	**Notes**
World	3769	3731	1.3		
Russia	206	92	−0.8		
United States	371	332	−4.5	62	
Canada	31	19	−4.2		
Australia	297	42	−1.9		
China	1747	1893	2.6	85	
India	294	424	6.3		
Europe	165	296	−2.2	85	

countries, with 13.7% per year achieved in China reflecting its rapid industrial growth.

- For countries well endowed with reserves, different ratios of consumption to production ranging from 0.3 to 1.6 were achieved in 2017 by Russia, the United States, Australia, China, and Qatar, reflecting different population sizes, manufacturing levels, and exports.
- Natural gas prices which reflect the wellhead source (e.g. in the United States, shale gas has a production cost advantage), whether liquefaction has occurred, and the extent of shipping, regasification, and distribution costs.

Considering crude oil, the highest rates of consumption growth from 2006 to 2017 have occurred in China and India (approximately 5% per year) whilst many countries have experienced small or even negative changes in consumption rates.

For coal, significant reductions in consumption rate have occurred from 2006 to 2017 in some countries (e.g. approximately 4% per year in the United States and Canada) whilst consumption has increased by 2.6% per year in China and 6.3% per year in India. Coal may be differentiated between thermal coal (used in steam and electricity generation) and metallurgical coal (used in converting iron ore to steel); for example, Australia exported 179 million tonnes of metallurgical coal and 203 million tonnes of thermal coal in 2017–18 [3].

Table 2.5 summarises some production and price data for some mineral commodities enabling chemical or mineral-based products. Useful data sources include

- US Geological Survey [21]
- British Geological Survey [22]
- World Bank Commodity Price Statistics [23, 24]

The World Bank report 'Commodity markets Outlook' October 2018 includes a graphical representation of monthly prices from January 2004 to January 2018, and annual prices from 1970 in both nominal and real terms, as well as price forecasts to 2030. The graphs provide a good indication of price volatility for the commodities reported. Historic data for production in different countries are also provided for certain commodities.

Table 2.6 summarises some production and price data for some key chemical products. Data for chemicals production in 2002 and 2012 are drawn from a report in the journal 'Chemical and Engineering News' [25] where production of a number of chemicals in China, the United States, Germany, Japan, South Korea, and Taiwan was reported for various product categories. The data indicate that whilst production over the period 2002–12 declined in some mature economies engaged in chemical processing, growth in China for important base inorganic and organic chemicals was very strong. World production data have been drawn from product reports by The Essential Chemical Industry online [26], which in turn has drawn on

Table 2.5 Production, price, and consumption patterns for some mineral commodities.

Commodity	Production 2017 (million tonnes)	Production 2016 (million tonnes)	Production 2012 (million tonnes)	Price 2017 (US$/t)	Consumption pattern United States 2016
Alumina					
World		118	94.1	355	
China		58.5	37.7		
Brazil		10.8	10.3		
Australia	20.6	20.7	20.9		
United States					70% to aluminium smelters; 30% to nonmetallurgical products, e.g. refractories
Aluminium					
World	59.7	58.8	49.4	2000	
China		31.9	23.5		
Russia		3.6	4.0		
Canada		3.2	2.8		
Australia	1.6	1.6	1.9		
United States		0.8	2.1		47% transportation; 20% packaging, 15% building, 8% electrical, 7% machinery
Bauxite					
World		289	257	29	
Brazil		39.2	35		
China		68	44		
Australia		83.5	76.3		90% converted to alumina; 10% to nonmetallurgical products

Table 2.5 Production, price, and consumption patterns for some mineral commodities—*cont'd*

Commodity	Production 2017 (million tonnes)	Production 2016 (million tonnes)	Production 2012 (million tonnes)	Price 2017 (US$/t)	Consumption pattern United States 2016
Phosphate rock					
World		276	220	90	
China		144	95.3		
Morocco		30	27.1		
United States		27.8	30.1		Majority processed to H_3PO_4 to make ammonium phosphate
Salt					
World		279.6	282.1	40	
China		63.1	69.1		
United States		37.2			80% used for chemical mainly Cl2, NaOH, manufacture
India		29.2	24.8		
Australia		11.5	13.6		
Sulphur					
					90% used for H_2SO_4 production
World total		83.3	82.5	80	
Pyrites, Frasch		8.6	8.7		
Recovered		69.4	68.8		
Sulphur ore		5.3	5.0		

a spectrum of sources supplemented by data from the US Geological Survey. The price data in Table 2.6 are based on a range of both historic and current data sources and should be regarded as approximate only; the many influences bearing on selling prices and their fluctuations must be carefully considered, as discussed under Section 2.5.3.

Table 2.6 Some indicative productions, production growths in various countries and indicative US prices for some common chemicals and polymers.

Country	Commodity	Production 2002 (kilotonnes)	Production 2012 (kilotonnes)	Production 2016 (million tonnes)	Price 2018 (US$/ tonne)
	Ammonia				
Japan		1450	1055		
Germany		2560	2696		
China				48	
United States		11,306	9131	9.8	350
World				146	
	Benzene				
Japan		4313	4214		
China		2131	26,405		
Germany		2106	1774		
South Korea		2582	4741		
United States				3.8	8050
World				46	
	Ethylene				
Japan		7152	6145		
China		5414	14,868		
Germany		4666	4897		
South Korea		5636	8075		
United States				26	1100
Middle East				20	
Asia Pacific				50	
Europe				20	
World				140	

Table 2.6 Some indicative productions, production growths in various countries and indicative US prices for some common chemicals and polymers—*cont'd*

Country	Commodity	Production 2002 (kilotonnes)	Production 2012 (kilotonnes)	Production 2016 (million tonnes)	Price 2018 (US$/tonne)
	Methanol				
China		2110	26,405		
United States				2	430
Middle East				9	
Asia Pacific				44	
Europe					
World				70	
	LDPE				
Japan		1789	1477		
United States		3647	3123		1350
World					
Europe					
	LLDPE				
United States		5139	6098		
World					
	HDPE				
South Korea		1871	2007		
Germany		1590	1834		
United States		7243	8046		
World					
	PVC				
Japan		2225	1332		
United States		6939	6944		
World				39	

Continued

Table 2.6 Some indicative productions, production growths in various countries and indicative US prices for some common chemicals and polymers—*cont'd*

Country	Commodity	Production 2002 (kilotonnes)	Production 2012 (kilotonnes)	Production 2016 (million tonnes)	Price 2018 (US$/ tonne)
	Phosphoric acid				
United States		10,125	8200	8.4	800
China				17	
World				43	
	Diammonium phosphate				
United States		10,825	6482		380
	Mono ammonium phosphate				
United States		4175	6482		
	Polypropylene				
Japan		2641	2390		
South Korea					
Taiwan					
China					
Germany		1755	1672		
United States		7691	7405		1300
	Polystyrene				
Japan		1837	1161		
South Korea					
Taiwan					
United States		3025	2473		

Table 2.6 Some indicative productions, production growths in various countries and indicative US prices for some common chemicals and polymers—*cont'd*

Country	Commodity	Production 2002 (kilotonnes)	Production 2012 (kilotonnes)	Production 2016 (million tonnes)	Price 2018 (US$/tonne)
	Sodium hydroxide				
China		8227	26,986		
Germany		3792	3490		
World				70	
Europe				10.7	
United States				11.4	400
	Sulphuric acid				
China		29,674	76,366		
Germany		2729	3918		
World				250	
United States				38	80
	Titanium dioxide				
World				5.7	
China				2.7	
Europe				1.3	
United States				1.3	
	Urea				
World				164	260
China				62	
India				23	
Middle East				20	
United States		4477	2475		

2.8 Selected case histories

Three case histories, one of a fuel commodity and two of chemical commodities, are now explored to illustrate or amplify some of the aspects previously discussed, and some of the challenges in market forecasting.

Case A. Natural gas

Natural gas is extracted from onshore and offshore resources, but requires processing to meet product specifications. Common impurities are water, hydrogen sulphide, and carbon dioxide. Wellhead gases are often associated with heavier hydrocarbons needing separation. Following purification, natural gas may be piped to onshore users or liquefied and shipped as liquefied natural gas (LNG) to other countries for regasification and use. Ethane is an important component of natural gas which is often extracted for feedstock in ethylene manufacture.

Fig. 2.5 indicates the key steps in bringing natural gas from the wellhead to a local or international market. Whilst wellhead gas composition may vary considerably [19], sales gas composition typically approximates 91% CH_4, 5.5% C_2H_6, 1.0% LPG, with nitrogen and carbon dioxide making up the balance.

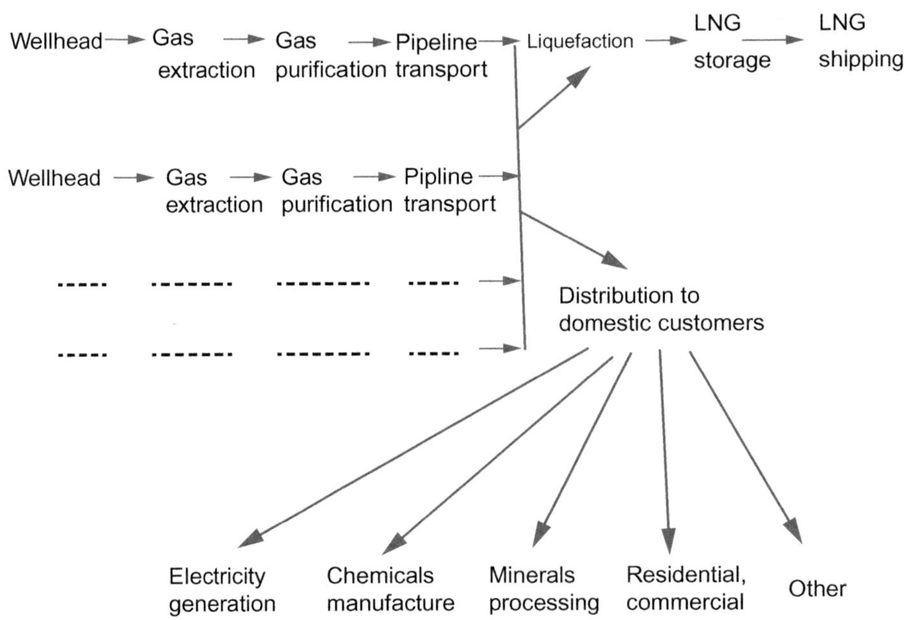

FIG. 2.5

Process system for gas supply to exports and/or domestic users.

Natural gas is a key commodity for applications throughout the economy encompassing:

- domestic heating, cooking, and hot water supply
- electricity generation for supply to the grid for industrial as well as commercial and domestic use
- dedicated industrial applications for steam and power generation and for high temperature duties in heating, drying, and calcining
- industrial use as a feedstock in a number of applications, including manufacture of
  - hydrogen
  - ammonia
  - methanol
  - ethylene, derived from ethane or propane

Fig. 2.6, which details the quantities of natural gas within Australia in 2012–13 for various applications [27], gives an indication of the diversity of natural gas uses.

Electricity generation from natural gas using a combined cycle has been a key use in many countries, offering major advantages over coal not only in terms of emissions of carbon dioxide, but also of particulates, sulphur dioxide, other pollutants, and solid waste generation through ash.

Open cycle generation, whilst less efficient than combined cycle generation, is commonly used for peak demand periods or for supplementing renewable sources.

Domestic electricity and gas consumption varies seasonally (e.g. air conditioning applications are greater in summer, whilst heating demands are greater in winter) and also in daily patterns (peak use tends to occur between early evening and late at night). Industrial demands can also vary corresponding to variations in plant capacity utilisation as well as variations reflecting the type of industry (e.g. day operation distinct from continuous shift operation).

Gas prices and their drivers vary between different regions of the world. Most LNG in Asia is sold under long-term contracts where the price of LNG is linked to the price of crude oil. The average price of LNG imported into Japan, the world's largest LNG purchaser, was approximately US$10/GJ in 2018, whilst crude oil prices were approximately $US72 per bbl.

A detailed account of natural gas exploration, extraction, transport and storage, usage, world trade and sales contracts is provided by Chandra [19]. Details of natural gas and liquefied natural gas projects completed or in progress in Western Australia reveal the long timelines typically incurred from resource discovery, through environmental and government approvals, through design studies, through construction to commissioning. A 10 year timeline for such events is not uncommon, during which there can be considerable volatility in LNG selling prices as well as construction costs. A useful summary of Western Australian gas projects, including details of capacities and capital expenditures, is provided in a Western Australian Government information sheet [28]. A separate fact sheet provides a detailed summary of milestones for the Gorgon project [29] which began LNG production in 2016.

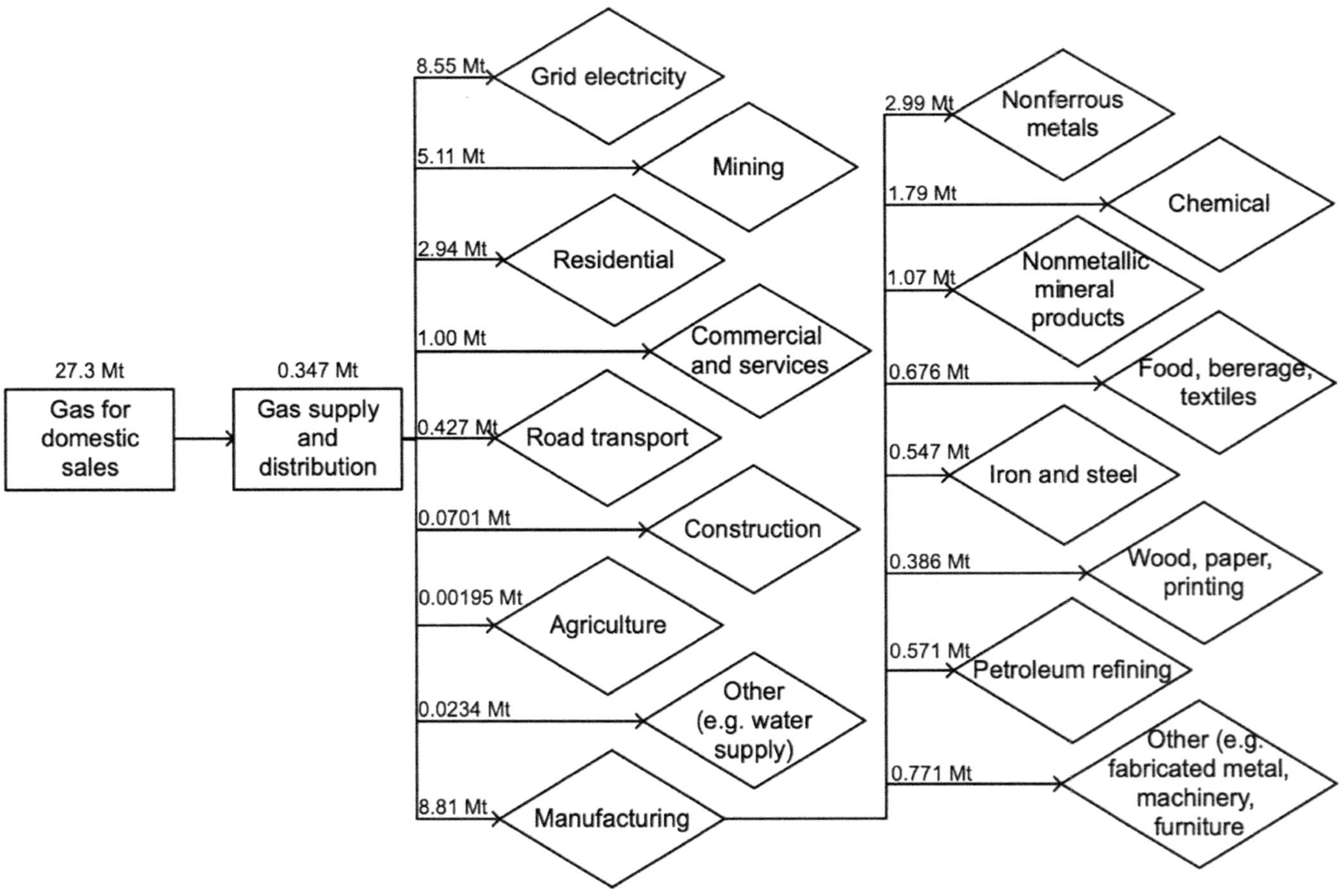

FIG. 2.6

Use of natural gas in Australia 2012–13 [27].

Case B. Sodium cyanide

A case product of interest, particularly in the Australian context, is sodium cyanide. Sodium cyanide is manufactured by producing hydrocyanic acid followed by neutralisation with caustic soda to produce aqueous sodium cyanide solution. The aqueous solution is then converted to solid sodium cyanide by evaporation, crystallisation, filtration (or centrifuging), drying, briquetting, and finally packaging. Sodium cyanide has a number of uses in electroplating, dyes, pharmaceuticals, and agriculture but its major use in Australia is for the extraction of gold from ores.

Prior to 1990, sodium cyanide had been imported into Australia. Hydrocyanic acid is readily available in many chemical economies as a co-product of acrylonitrile, manufactured from ammonia and propylene; acrylonitrile is manufactured from these raw materials in Japan, Europe, and the United States but not hitherto in Australia. Thus in Australia, hydrocyanic acid had to be manufactured to enable subsequent conversion to sodium cyanide. The most common commercial method available was from methane and ammonia (both derived from natural gas) using the Andrussow process.

Imports of sodium cyanide into Australia in the mid to late 1970s were approximately 2000 t/annum but higher world gold prices have led to a substantial increase in the quantity of gold mined and extracted in Australia since 1980. Imports of sodium cyanide to Australia increased substantially from 2000 t in 1975 to 52,000 t in 1988. Three Australian plants came on stream around 1990 to manufacture sodium cyanide, two making a solid product and the third a liquid product. Two are still in operation and have been expanded in capacity since their initial construction. In 2018, sodium cyanide was manufactured as both liquid and solid products by Orica at Gladstone with a total capacity of 95,000 tonnes per annum NaCN, and by Australian Gold Reagents at Kwinana with a total capacity of 78,000 tonnes per annum. Some 50% of production is now used locally and 50% exported.

In retrospect, demand for sodium cyanide in 1990 would have been most difficult to forecast in 1975 and the 1980s. The world price of gold has been volatile, and from 1975 to 1988 approximately doubled in real terms (in US $/oz). During the same period, world production of gold increased by 30% whilst the Australian share of world gold production increased from 1.4% to around 10%. Gold mine production was approximately 3300 tonnes globally in 2018 with 303 tonnes of gold produced in Australia.

The manufacture, transport, and use of sodium cyanide has environmental and safety risks [30]. Liquid sodium cyanide transport is restricted because of the safety and environmental implications of accidental spillage in transportation. Hence there has been interest in the potential for smaller-scale manufacture of liquid sodium cyanide at the point of use in remote gold mining centres in Australia [31]. Such initiatives, whilst suffering from penalties from smaller-scale manufacture of liquid sodium cyanide, would avoid the additional costs of evaporation, crystallisation, drying, and packaging required in the manufacture of the solid product, as well as transport from the manufacturing site to the user.

Recently, a novel process has been researched using an ammonia-free process suitable for mine site production, and an extensive pilot plant trial is currently being explored in Tasmania [32, 33]. Research has also been undertaken in Australia by CSIRO [34] into the potential for using thiosulphate as an alternative chemical for extraction. The proposed process is being trialled in partnership with Eco Minerals Research at a small gold mine in Menzies, Western Australia.

The case of sodium cyanide in Australia highlights a number of important influences on market development:

- the price, demand, and availability of mineral resources (in this case gold) in a particular region or country
- the availability of suitable feedstocks for manufacture (in this case natural gas and caustic soda have been available for industrial use in Australia)
- capability for growth in manufacturing capacity
- incentives for smaller-scale manufacture of toxic chemicals at the point of use
- incentives for alternative technologies and potential impact on existing products and manufacturers

Case C. Ethylene

Ethylene is a dominant petrochemical product with reported global production in 2018 of some 150 million tonnes per year. It is of interest because of the evolution in its process technology, the growth in scale of its process plants, the dependence on hydrocarbons in its production both for feedstock and utilities, and its role as a feedstock for many downstream chemical products of commercial importance. Ethylene is also important because of its sensitivity to global economics and energy supply; steam cracking for ethylene production has been identified as the most energy consuming process in the chemical industry [35]. Ethylene manufacture is currently celebrating a centenary since the initial process for its manufacture was developed; some historic and current aspects of its production are now reviewed.

Production origins in the United States of America

In the 1920s, a process for cracking ethane/propane mixtures and separation to produce ethylene was developed within the Union Carbide Corporation which founded its first petrochemical industry in Charleston, West Virginia. The period 1935–50 saw some important petrochemical processes emerge from research and development to commercial realisation; this emergence had an impact on demand for ethylene. Examples included:

- 1936—Vinyl chloride/vinyl acetate copolymers by Union Carbide Corporation.
- 1937—First direct oxidation of ethylene to ethylene oxide by Union Carbide Corporation.
- 1938—First large-scale styrene and butadiene plants (Dow Chemical Co.).
- 1943—I.C.I. polyethylene process operated in the United States (Du Pont and Union Carbide Corp.).
- 1948—First direct hydration plant for ethanol produced from ethylene (Shell Chemical Co.).

During World War II there was a rapid increase in demand for ethylene derivatives, and ethylene consumption grew to 130,000 tons/annum in 1946. Average growth rate approximated 16%/annum from 1940 to 1955, and 11%/annum from 1955 to 1973. Growth from 1973 to the early 1980s was severely retarded by sharp price rises in crude oil and hence in hydrocarbon feedstocks for ethylene manufacture. Growth was later restored however with an average growth rate of 5% per annum achieved from 1985 to 1995.

Feedstocks for ethylene production

The petrochemical industry developed in the 1960s and early 1970s on the basis of surplus feedstocks for olefins, predominantly ethane from natural gas in the United States, and naphtha from refining in Western Europe. The oil shocks of 1973/74 and 1979 had a major impact on feedstock sources. Europe started to diversify away from naphtha towards lighter feedstocks, principally LPG and ethane, whilst other regions, particularly the Middle East, began to develop an ethylene industry based on low value ethane. In 1994, Europe and Japan still relied heavily on naphtha, whilst the Middle East, Canada, and the United States had a predominant reliance on ethane. In the US in 2017, ethane typically accounted for 73% and LPG 16% of ethylene feedstock with LPG accounting for 16%. In contrast, China, which has undergone a rapid increase in ethylene production during this century, has relied heavily on naphtha for feedstock.

Co-products increase in quantity and commercial importance as the feedstock becomes heavier from ethane (where 125 tonnes of ethane produce 100 tonnes of ethylene and around 3 tonnes of propylene) to naphtha where 330 tonnes of feedstock are required to produce 100 tonnes of ethylene plus typically 52 tonnes propylene, 14 tonnes butadiene, and 22 tonnes of benzene. Propylene derivatives include polypropylene, acrylonitrile, propylene oxide, phenol, isopropanol, and acrylic acid.

Ethylene derivatives

Polyethylene, encompassing low-density (LDPE), linear low-density (LLDPE), and high-density (HDPE), accounts for around 60% of total ethylene demand. LDPE and LLDPE tend to be used for film in packaging, with HDPE used for more diverse applications. The next largest use is ethylene oxide used to make ethylene glycol, followed by vinyl chloride monomer used to make PVC.

Trends in demand, production cost, and selling price

World forecasts tended to underestimate the demand for ethylene in the 1960s [16] and overestimate demand in the late 1970s and early 1980s [35]. Price trends for ethylene feedstocks have been closely linked to trends in natural gas and crude oil prices. In the 1960s and early 1970s, crude oil prices remained steady in nominal terms at around $2 US/bbl leading to cheap feedstocks. With sharp price increases to $12 US/bbl in 1974 and to $28 US/bbl in 1980, ethylene costs soared and demand slowed. Some new manufacturing capacity was already under construction in 1974. Many forecasts since 1974 up to the early 1980s underestimated the dampening in demand caused by cost increases, being guided by the long and impressive record of growth in the 1950s, 1960s, and early 1970s. Superimposed on the oil price effect in the later 1980s was the pronounced cyclic fluctuation reflecting the changing imbalance between supply and demand. Historic trends in the real prices of crude oil and ethylene over the period 1960–95 are reported in Chapter 2 of the first edition of this book [36]. More recent trends are reported in a range of publications for example [37, 38]. In 2016, the US ethylene prices have been reported at around US$660/tonne whilst global ethylene prices were reported at around $US1000/tonne [38].

Current production and scale of new plants

In 2018, world production reached 154 million tonnes per year, a doubling of that achieved 25 years earlier, with 37% in Asia Pacific, 18% in the United States, 15% in Europe, and 15% in the Middle East. Major growth has been achieved in China. US Gulf Coast capacity has increased over the period 2016–19 by a further 6 million tonnes per year. In 2017, Dow Dupont commissioned a 1.5 million tonne/year ethylene plant benefitting from competitively priced shale gas in the United States. Similar sized plants were scheduled for commissioning in the United States by Chevron Phillips and Exxon Mobil [38].

References

[1] R.F. de la Mare, Parameters affecting capital investment, Chem. Eng. 296 (1975) 227–230. 247.

[2] D.H. Allen, Economic Evaluation of Projects, third ed., IChemE, Rugby, UK, 1991.

[3] Australian Government, Resources and Energy Quarterly, June, www.industry.gov.au/req, 2019.

[4] C. Kline, Maximising profits in chemicals, Chemtech (February) (1976) 110–117.

[5] J. Wei, T.W.F. Russell, M.W. Swartzlander, The Structure of the Chemical Processing Industries, McGraw-Hill, New York, 1979.

[6] J.R. Couper, O.T. Beasley, W.R. Penney, The Chemical Process Industries Infrastructure, Marcel Dekker, New York, 2001.

[7] R. Smith, Chemical Process Design and Integration, Wiley, Chichester, UK, 2005.

[8] W. Seider, J. Seader, D. Lewin, S. Widago, Product and Process Design Principles. Synthesis, Analysis and Evaluation, third ed., John Wiley and Sons, New York, 2009.

[9] M. Sigurdsson, D.F. Rudd, Trends in the international production and trade in petrochemicals, Chem. Eng. Commun. 66 (1) (1988) 125–168.

[10] D. Seddon, Petrochemical Economics. Technology Selection in a Carbon Constrained World, Imperial College Press, London, UK, 2010.

[11] G.M. Wells, Handbook of Petrochemicals and Processes, second ed., Aldershot, UK, 1999.

[12] Kirk-Othmer, Encyclopaedia of Chemical Technology, fifth ed., 2005, John Wiley, New York.

[13] Ullmann's Encyclopaedia of Industrial Chemistry, 2012, Electronic Release, Wiley, Weinheim, Germany.

[14] J.H. Gary, G.E. Handwerk, M.J. Kaiser, Petroleum Refining' Technology and Economics, fifth ed.CRC Press, Boca Raton, FL, 2007

[15] I.M.M. Aus, Australasian Mining and Metallurgical Operating Practices, third ed., Carlton, VIC, Australia, 2013.
[16] D.S. Remer, C. Jorgens, Ethylene economics and production forecasting in a changing environment, Eng. Process. Econ. 3 (1978) 267–278.
[17] Fischedick, et al., 'Industry' (Chapter 10) in Climate Change: Mitigation of Climate Change. Contribution of Working Group III to the Fifth Assessment Report of the Intergovernmental Panel on Climate Change, Cambridge University Press, Cambridge, UK, 2014.
[18] D.F. Rudd, C.C. Watson, Strategy of Process Engineering, John Wiley and Sons, New York, 1968.
[19] V. Chandra, Fundamentals of Natural Gas. An International Perspective, (2006) PennWell.
[20] BP Statistical Review of World Energy, 67th ed., July, 2018.
[21] USGS, Mineral Commodity Summaries, US Geological Survey, Virginia, 2017.
[22] BGS, World Minerals Production 2012-2016, British Geological Survey, Nottingham, UK, 2018.
[23] World Bank Group, Commodity Markets Outlook, October, 2018, Washington, DC.
[24] World Bank Commodity Price Forecast, October, 2018, Washington, DC.
[25] Facts and figures of the chemical industry, Chem. Eng. News (July 1) (2013) 41–48.
[26] The Centre for Industry Education Collaboration, The Essential Chemical Industry Online. http:/www.essentialchemicalindustry.org/.
[27] Bureau of Resources and Energy Economics (BREE), 2013 Australian Energy Statistics, CanberraAustralian Government, Australia, 2013.
[28] WA Liquefied Natural Gas Industry Profile, Government of Western Australia, February, 2018.
[29] Chevron Corporation, Gorgon Project Milestones Summary, Fact Sheet. https://australia.chevron.com/-/media/australia/our-businesses/documents/Gorgon-Project-Milestones-Fact-Sheet.pdf, 2017.
[30] NICNAS, Priority Existing Chemical Assessment Report No. 31 on Sodium Cyanide, Australian Government, Canberra, 2010.
[31] D. Brennan, Sustainable Process Engineering, Chapter 15, Pan Stanford, Singapore, 2013.
[32] Pradeep S, Duckworth, G. Dunks, C., Rudolph, V., Paper No. 1240, Proceedings of Chemeca, Perth, Western Australia, 2014
[33] G. Duckworth, P. Shukla, Mine site cyanides, The Chemical Engineer March, 2015, 36–39.
[34] R. Nicoll, Gold Strike with thiosulphate, CSIRO J. - Resourceful (7) (2015).
[35] G.S.G. Beveridge, The Institution of Chemical Engineers Presidential Address, IChemE, UK, 1984-85.
[36] D. Brennan, Process Industry Economics. An International Perspective, Institution of Chemical Engineers, Rugby, UK, 1998.
[37] S. Lewandowski, Ethylene global, in: Presentation at Asia Chemical Conference, November, 2016.
[38] D. Lippe, Uncertainty looms for US olefins industry, Oil Gas J. March, 2018, 64–71.

CHAPTER

Capital cost estimation

3

For which of you, desiring to build a tower, does not first sit down and count the cost, whether he has enough to complete it?

St. Luke Chapter 14 v 28

Chapter outline

Process Industry Economics. https://doi.org/10.1016/B978-0-12-819466-9.00003-1

3.1 Capital requirements and sources

Capital investment can be regarded philosophically as deferred consumption of wealth in the expectation of greater wealth or well-being in the future. Capital investment in the process industries is required for entirely new plants, for modifications to existing plants, and even for minor changes in order to sustain a plant during its operating life. Investment into new projects can involve massive sums; world scale alumina refineries, aluminium smelters, ethylene plants, or pulp mills, for example each require investments exceeding US$1 billion. Plant modifications or revamps are frequently necessary to accommodate changes in technology, feedstock or product quality specifications, to reduce operating costs, to comply with changing safety and environmental legislation, and to enable increased production capacity. Revamp projects typically involve expenditures in the order of tens of millions of dollars. Investment required merely to sustain an existing plant is somewhat less, but nevertheless represents an ongoing commitment.

Capital requirements for process plants can be classified under the following categories:

- Land, where the investment is potentially recoverable at the end of the project life, though expenditure may be incurred in site remediation;
- Fixed capital investment into plant and buildings, which generally has negligible recovery value (relative to initial investment) at the end of the project life;
- Start-up capital, which is spent during plant commissioning and is nonrecoverable;
- Working capital, which represents investment into stocks of feedstocks and products essential for plant operability and extended credit to customers, all of which is essentially recoverable at the end of the project life.

Capital may be derived from a number of sources, but these may be broadly classified as either equity or loan sources. Equity funds are funds owned by a company derived from shareholders' funds and from retained earnings. Retained earnings may include provisions for depreciation of capital assets. Loan funds are borrowed from external bodies such as banks or insurance companies and must be repaid with interest. Equity funds do not require interest payments (though dividends are paid to shareholders) but have earning capability through investment opportunities. Such investment opportunities include process industry projects and also quite unrelated areas. Thus equity funds have an 'opportunity cost' equal to their earning capability through alternative investment. Capital funds from whatever source have a cost which is related to perceived risk. Various capital sources may have different interest rates or opportunity costs, but these can be combined to give a weighted interest rate or so-called 'cost of capital', conventionally expressed as % per annum.

3.2 Land costs

Land for a process industry plant may be apportioned from an existing site owned by the company concerned, or may need to be purchased as a new site. If the land is apportioned from an existing site there is no corresponding cash flow, but a nominal cash transfer may be charged to the project. Such a cash transfer can be pooled with similar cash transfers from other projects to assist the purchase of other sites when existing land sites become fully utilised or are unsuitable for use.

If a new site is purchased, the capital cost will be a function of the size and location of the site involved. Such costs vary widely with individual locations but are generally small in magnitude compared with fixed capital costs of plant and buildings constructed on the land. Some desirable characteristics of land for process industry use are that it should be flat, well drained, have soil of good load-bearing characteristics, be a safe distance from urban development, have good access to road, rail, and sea transport, and have access to raw materials, utilities, labour, and markets. Not all of these desirable characteristics can always be satisfactorily met, and compromise may be necessary. An additional important consideration is that it should be secured from the influence of natural calamities such as earthquakes or floods.

Site preparation and development, and the obtaining of clearances from various government authorities are usually more time consuming for a new site than for an existing site. New sites are often termed '***greenfield***' sites while existing sites are termed '***brownfield***' sites. Brownfield sites may, however, present a number of constraints to new plant construction arising from existing plant operations, leading to additional construction cost burdens.

3.3 Fixed capital investment

For process plants, fixed capital investment can be subdivided into:

(i) inside battery limits (IBL);
(ii) outside battery limits (OBL) or 'off-sites'.

'Battery limits' may be defined as a geographic boundary, real or imaginary, around the processing plant, which converts raw material to finished product. It is common to account for the cost of the processing plant in terms of inside battery limits (IBL) costs, which are governed by the plant production capacity, the qualities of the raw materials and products, and the process technology employed. The efficiency of raw materials conversion and of energy and utilities consumptions, as well as safety and environmental standards, can exert considerable influence on the design of process and plant, and thus on IBL costs.

3.3.1 Outside battery limits investment

Outside battery limits (OBL) also referred to as 'off-sites' costs include:

- storage and handling facilities for raw materials and finished products;
- utilities generation facilities including plant for steam and electricity generation, cooling towers for cooling water supply, plant for purge gas and compressed air supply, and effluent treatment facilities;
- buildings and service facilities for process plants incorporating laboratories, workshops, offices, warehouses, cafeteria, and medical facilities.

The distinction between IBL and OBL cost areas is illustrated in Fig. 3.1.

The design and costs of storages depend on the quantities and properties of the raw materials and products stored; relevant properties include whether materials are stored as gases under pressure, as liquefied gases, as corrosive or inflammable liquids, and so on. Consumption of raw materials depends on the process technology and capacity of the IBL plant. However, storage quantities are often far more

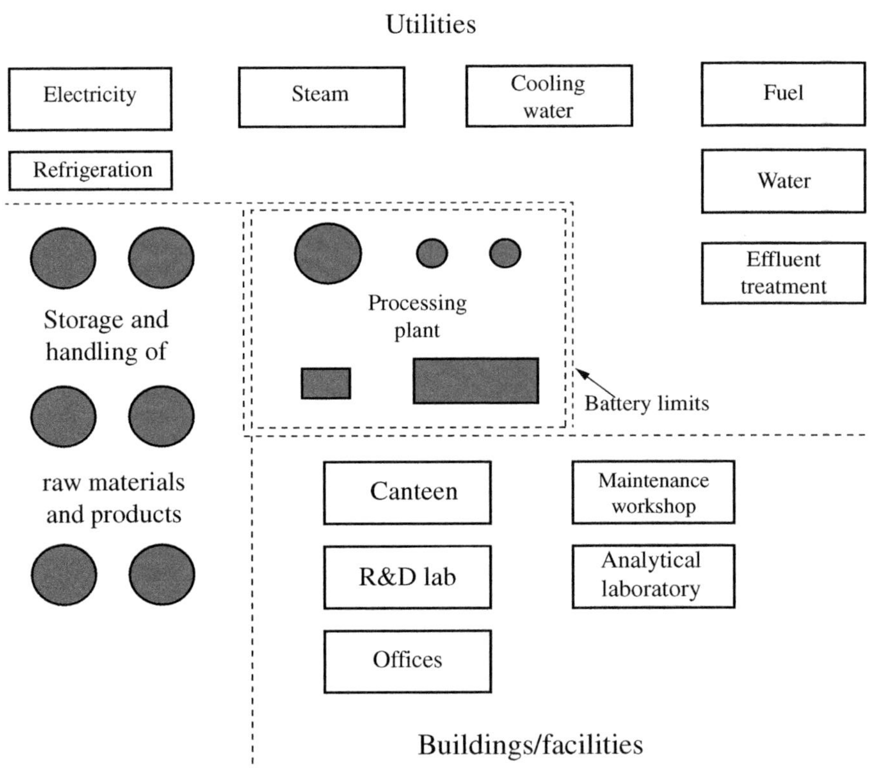

FIG. 3.1

Inside and outside battery limits.

dependent on the way in which the project is implemented, particularly in relation to transport of raw materials and products, and its integration with other manufacturing facilities. For example, if the raw material is supplied as the product from another plant on the same site it can be piped with minimal intermediate storage. If however the raw material is imported by ship, storage requirements could be substantial depending on shipping frequency. Similarly, if the product from the plant is piped to an adjacent plant on the same site, storage requirements may be minimal; if however, the product is to be shipped or transported by rail or road, storage requirements may be substantial.

Consumption of utilities also depends on the process technology and plant capacity. The decision to either generate or purchase a utility has obvious effects on capital investment into utilities. On a site complex, if utilities generation facilities are shared with other manufacturing plants, scale economies should result. Surplus steam may also be available on a site complex from a neighbouring plant.

Buildings and service facilities depend to some extent on the industrial facility concerned. Some savings can be expected if buildings or service facilities are shared between multiple process plants. Note that some buildings such as control rooms or compressor houses are located within plant boundaries and are part of IBL capital.

For a given IBL capital investment reflecting a particular plant capacity and process technology, the magnitude of the related OBL capital can thus vary considerably depending on where the project is located and how it is implemented. OBL capital can typically range between 5% and 100% of IBL capital depending on the process technology, site location, and project implementation. ***Hence rules of thumb which estimate OBL capital as a percentage of IBL capital, even for a given technology, must be approached with caution.***

Note that the terminology here is somewhat imperfect. 'Off-sites' are usually outside the battery limits of the processing plant but usually on the same (larger) site. In terms of site layout, off-sites for a particular plant may be grouped with off-site facilities for other plants especially in the case of storage facilities, or may be more isolated. The term 'services' is often used to mean utilities; the term 'plant service facilities' is often used to include buildings. ***There is need to ensure that the terms adopted in a project's capital cost estimate are clearly defined and understood by the various parties participating in the project.***

While the distinction between inside and outside battery limits capital is helpful, it is possible that some cost estimators may not adopt the distinction. It is thus important that the scope of cost categories be well defined by the estimator and scrutinised by the analyst to avoid any double counting of costs associated with storages, utilities, and buildings.

3.3.2 Inside battery limits investment

Considering IBL plant costs, equipment items can be readily identified from well-defined flowsheets at the design stage, or from the physical plant itself at the constructed stage. For a project, they will be clearly identified on an equipment list.

Sometimes referred to as 'main plant items', these include reactors, heat exchangers, filters, pumps, compressors, separation columns, and so on. Their purchased costs including the costs of associated drives (usually electric motors or steam turbines) are summed to give the purchased equipment cost (PEC). The PEC is an identifiable and important component of plant cost. The process plant cost (PPC) is commonly related to the PEC by a 'Lang' factor (f_L) according to the expression

$$\text{PPC} = f_L * [\text{PEC}] \tag{3.1}$$

The value of the Lang factor for most process plants is in the range of 3–4, reflecting the other costs contributing to plant cost. These other costs include:

- site preparation and provision of drainage, grading, roads, and access ways;
- foundations and structures for equipment items;
- installation of equipment on site involving some materials but mainly labour charges;
- specialised buildings such as control rooms, and those for housing certain equipment items or sections of plant;
- piping;
- instrumentation;
- electrics;
- painting and lagging;
- overheads charges for design, equipment and materials procurement, project management, construction management, and commissioning;
- where applicable, capital charges associated with the purchase of technology.

This division of cost categories along with purchased equipment costs is frequently used in the preparation and presentation of cost estimates. Most cost categories, in particular piping, electrics, and instrumentation, are frequently further divided into materials and labour components.

A number of methods are available for estimating the capital costs of process plants and find application for different purposes such as:

- economic planning by companies or government departments related to the industry;
- selection and management of research and development projects;
- feasibility studies for projects;
- optimising the design of a process or plant;
- sanction of a capital expenditure proposal by the company intending to operate the proposed plant;
- tendering for a contract by a design office to manage the design, construction, and commissioning of a plant;
- control of capital expenditure during the design and construction stages of a project.

These methods require different degrees of detail and engineering expertise for data input, require different levels of work hours, incur different levels of cost, and result in different levels of accuracy. There is thus an important relationship between the cost of preparing an estimate and its likely accuracy.

The nominal accuracy required of a sanction estimate is normally ±10%, and the work required to enable its preparation may amount to 2%–5% of the total cost of the project. For this level of accuracy, the site would be chosen, process and engineering flowsheets developed, equipment items specified and costed, preliminary site and plant layouts developed, and preliminary utility diagrams prepared. Some preliminary design of buildings, structures, civils, instrumentation and electrics would also be completed, with associated materials requirements for piping, electrics, instruments, concrete, and steel work. Estimates would also be made of personnel and time requirements for project implementation.

For a detailed estimate typically used by a contractor in tendering for project execution, the accuracy might be ±5%, requiring a more detailed level of engineering design and estimating, and implying a cost of preparing the estimate in excess of 5% of the project cost.

For an initial feasibility study involving consideration of alternative sites or process routes, a somewhat lower level of accuracy, say ±30% might be satisfactory, representing a cost of around 0.2% of the total project cost. Such an estimate might rely extensively on cost data for previous plants using similar technology, assisted by more detailed analysis in higher cost areas.

It is common practice for estimators to add a 'contingency' category to the estimated project costs to cover those unforeseen costs which are always present in a project, and which reflect uncertainties in design and project implementation. Contingency allowance might be ±30% for a feasibility study, and ±10% for a detailed estimate.

AACE International (association or the Advancement of Cost Engineering) has developed a Cost Estimate Classification System [1] outlining five classes of cost estimates for engineering, procurement and construction (EPC) work. Many process engineering contracting firms adopt this classification. Each class of cost estimates has distinguishing features in terms of:

- the maturity level of project definition deliverables expressed as a percentage of complete definition;
- the 'end usage' or typical purpose of the estimate;
- the methodology employed in generating the estimate;
- the expected accuracy range, with typical variation in low and high ranges.

The typical purpose or 'end usage' related to each class are broadly classified as:

- Class 5: Concept screening
- Class 4: Study or feasibility
- Class 3: Budget authorisation or control
- Class 2: Control or bid/tender
- Class 1: Check estimate or bid/tender

From a chemical engineering perspective, the contributions of other engineering disciplines and the benefit of previous project experience are essential inputs into realistic cost estimates. The flow of information from the chemical engineer as the originator of the process and plant concepts, to other engineering functions and to those responsible for producing the estimate, are important elements in the development of the capital cost estimate.

In the evaluation of alternative processes at the initial design stage, or in early stages of process flowsheet development for an evolving process design, it is important to incorporate capital cost considerations. Often this is done without the availability of detailed engineering and estimating expertise. In these cases, an understanding of the structure of plant costs and an ability to make approximate estimates, become important skills for the chemical engineer.

Two of the more common methods of overall plant cost estimation are discussed under Sections 3.4 and 3.6. The first involves extrapolation from known plant cost data and has a typical accuracy of ±50%, while the second involves applying factors to purchased equipment costs and has a potential accuracy of ±30%. In assigning nominal accuracies to estimates, it should be appreciated that there is usually a greater probability of underestimating than overestimating capital requirements for projects, derived from unforeseen details. Thus the bulk of experience would indicate accuracies more realistically of +A/−B% where A > B.

3.4 Plant cost estimation by extrapolation from known plant cost data

An approximate estimate of the IBL fixed capital cost of a proposed plant can be made from knowledge of an achieved cost or previously estimated cost of a reference plant of similar scope, and using the same process technology. Adjustments for inflation will nearly always be necessary, and adjustments for differences in capacity or location are often required. A generalised relationship for making such adjustments may be expressed as

$$I_p/I_r = (Q_p/Q_r)^b * F_p/F_r * L \tag{3.2}$$

where

I = fixed capital investment
Q = production capacity of plant
F = inflation index
L = location factor
b = an exponent
p denotes the proposed plant
r denotes the reference plant

In using reference plant cost data, it must be acknowledged that there are many influences on the achieved cost of a plant. These influences acting in combination can lead to considerable scatter in plant cost data. The influences include:

- Effects of various business conditions for contractors, fabricators, and construction labour. Business conditions include the supply-demand balance for services, time schedules for work completion, financial value of a particular contract, and expertise of management.
- Scope of plant design, including off-sites. This may vary with different arrangements for feedstock or utility supply, or for different degrees of co-product or heat recovery.
- Effects of safety, occupational health, and environmental requirements on plant design details; such requirements vary for different sites and different countries and is often influenced by government policy. Even within countries, government policy and investment incentives may vary between national, state, and regional governments.
- Productivity of personnel, especially construction labour.
- Location effects of the plant site, both within a country and between countries.
- Effects of inflation.

The approach is more reliable when restricted to IBL investment since OBL investment depends on how the project has been implemented and is likely to be much more variable from one project to another. The approach is now discussed under the separate aspects of capacity, inflation, and location adjustments.

3.4.1 Capacity adjustment

Capacity adjustments are usually made on the basis of the relationship

$$I_p/I_r = (Q_p/Q_r)^b \tag{3.3}$$

where the exponent b depends on the flowsheet structure of the plant. Where capacity is achieved by multiple equipment items in parallel, the value of b approaches 1. However, in all plants, there are some costs which are largely independent of capacity, for example control room costs and design costs. Further, because of scale economy pressures, few plants are parallel streamed throughout. Thus for plants which are nominally parallel stream

$b = 0.8\text{–}0.9$

For plants which are single stream

$b = 0.5\text{–}0.6$

These values, along with the relationship itself, are based on a large volume of accumulated, empirical experience, but should only be regarded as an approximation.

For plants which are mixed parallel/single stream, values of b are intermediate between those for single and parallel stream plants. Allen and Page [2] have proposed a relationship for relating the exponent b for a process plant to the exponent b for individual equipment items

$$b = \sum_{i=1}^{J} c_i b_i \Big/ \sum_{i=1}^{J} c_i \tag{3.4}$$

where

J = number of equipment items (excluding multiples of a given item)
c = purchased cost of main equipment items of base capacity.

Where multiples of a given equipment item are used in parallel to achieve the required capacity, the exponent b is assumed to approximate a value of 1.

A simpler way of evaluating b for a mixed parallel/single stream plant is to weight typical values of b based on the capital distribution for entirely single streamed or parallel streamed sections of the plant. For example, the IBL capital distribution for an ethylene plant is approximately 30% for multiple reactors and associated waste heat boilers and 70% for single stream purification; thus an exponent of 0.3 * 0.9+0.6 * 0.7 or 0.69 is predicted. The IBL capital distribution for a chlor-alkali plant approximates 50% for multiple cells and rectifiers and 50% for single stream brine treatment and purification of products; thus an exponent of 0.5 * 0.9+ 0.5 * 0.6 or 0.75 is predicted. Both these predicted values agree with reliable published data.

There is evidence that the values of exponent b are not always constant over the entire plant capacity range. Miller [3] claimed that the exponent may increase from 0.5 to 0.9 over the available capacity range, while Taylor and Craven [4] based on UK experience reported exponents for continuously operating single stream plants increasing from 0.3 for capacities of 500–1000 t/year to 0.6 for capacities of 100,000–200,000 t/year. Such evidence provides a cautionary warning in estimating costs of small plants, common for new technologies and for cases where capacity is matched to domestic markets in developing or low population regions. Miller [3] also reported that scale economies can be different for IBL and OBL capital, and recommended that separate evaluations be made for each category of capital.

3.4.2 Location factors

Location factor denotes the ratio of the cost of a plant built in the proposed location, to the cost of an identical plant built in the reference plant location, both costs being expressed in the same currency. The bulk of published plant cost data is from the United States, and frequently this data is used as a guide to costs of building a similar plant in another country. The location factor for Canada compared with the United States, for example may be defined by the expression:

$$\text{Location factor} = \frac{\text{Cost of plant in Canada (US\$)}}{\text{Cost of equivalent plant in USA (US\$)}}$$
$$= \frac{\text{Cost of plant in Canada (Can \$)}}{\text{Cost of equivalent plant in USA (US\$)}} * \text{Value of Can \$ in US\$} \quad (3.5)$$

Note that in order to estimate the costs in the currency of the proposed location, the currency exchange rate must be used. By monitoring trends in exchange rates and inflation rates for any two countries over time, it can be appreciated that the location factor between two countries can change with time. Location factors depend both on relative labour costs and relative labour productivities which differ for different countries and which can also change over time. Location factors published for Saudi Arabia relative to US Gulf Coast [5], for example ranged between 1.48 and 1.98 over the period 1974–1985 (see Table 3.1); location factors quoted from various sources for Australia vs United States ranged between 0.83 and 1.3 over the period 1970–1990 [6]. Thus a location factor should ideally be qualified by reference to a particular year.

Location factors can also be different for plants of different process technologies due to differing contributions of buildings, site construction, overseas components, freight, and overheads to plant costs.

Martin [7] reported location factors as of first quarter 1994 for several countries for a hypothetical fine chemicals/pharmaceuticals plant with a total installed cost of approximately US$35 million (US Gulf Coast). The assumed cost breakdown for the reference plant comprised:

- 15% engineering
- 37% materials
- 48% construction.

Based on estimates of relative man-hour rates and relative man-hours required for engineering and construction costs, estimates were made of relative project costs. Costs relative to the United States for a selected group of these countries were estimated as

- United States 1.0
- United Kingdom 1.13
- Indonesia 0.72
- India 0.81.

While the cost breakdowns for process plants differ for different technologies and relative costs vary over time, the approach provides a useful insight into how international cost comparisons might be made.

Indicative location factors for process plants are available under commercial terms from a number of companies, for example Compass International [8].

Further challenges arise in establishing location factors for different regions within one country. For example, within the United States different capital costs of process plants may be experienced for Gulf Coast, East Coast, West Coast, and

Table 3.1 Saudi Arabia location factors relative to US Gulf Coast [5].

Year	1974	75	76	77	78	79	80	81	82	83	84	85
Location factor	1.73	1.66	1.63	1.69	1.98	1.85	1.74	1.60	1.48	1.56	1.57	1.53

so on. Gary et al. [9] have reported relative plant costs for refinery hydrocarbon processing on a 2005 basis for various US locations; these range from 1.0 on US Gulf Coast through 1.3 at Chicago, 1.5 at Philadelphia, to as high as 2–3 at Alaska. Main factors contributing to these variations within the United States are stated as climate and its effect on design requirements and construction conditions; local regulations, codes and taxes; and availability and productivity of construction labour.

Kjar [10] has provided some location factor estimates within Australia. Factors range from 1.0 in major capital cities to 1.1–1.4 in major resource areas in or near settled communities to 1.4–1.8 in remote areas such as Pilbara, Kimberleys, and North Queensland. Considerations include the remoteness of some locations with additional delivery costs for materials, and higher labour costs reflecting the need to compensate for remoteness, sometimes including a fly-in, fly-out basis for construction employees.

While there are undoubtedly location differences in components of plant costs for different regions of Europe, these are generally minor in terms of overall plant costs for most industrial locations of Europe.

3.4.3 Plant cost inflation indices

Because of the recurring needs to update historic capital costs or to compare capital costs from different time periods, cost inflation indices are required which reflect the changes in capital costs of plants with time. A number of such indices have been developed for different countries and have in common a weighting of various contributions which collectively make up the cost of a process plant.

For the United States, the Chemical Engineering Plant Cost Indices published in the journal *Chemical Engineering* have provided detailed compositions reflecting the movements in equipment, plant, and construction costs. A similar cost index was reported for many years as the Nelson Refinery Cost Index published in the *Oil and Gas* journal, though the publication of this index ceased in 2017. For other countries the availability of statistics is rather more limited; in such cases, a composite of only two indices representing the costs of materials and labour has commonly been adopted. For example in Australia, a composite index has been used by the author based on 50% contribution from average weekly earnings, and 50% materials price index. The materials price index was initially that for materials used in buildings other than houses and was replaced by the producer price index providing similar data in 2005. Both components are published by the Australian Bureau of Statistics. The Chemical Engineering Plant Cost Index, the Nelson Refinery Cost Index, and the Australian Plant Cost Index (as estimated by the author) are listed in Appendices 3 and 4.

Some important constraints on the use of inflation indices in updating plant costs must be emphasised. Inflation indices in almost all cases make no allowance for changes in:

- technology;
- labour productivity;
- scope of plant design.

During the 1970s, a period of widespread high inflation in most developed countries, the cost of plants escalated much more than predicted by the plant cost inflation indices. This can be attributed to

- more sophisticated, capital intensive designs of plants to provide improved safety and environmental standards, and improved energy and feedstock efficiencies;
- rapid escalation of site labour costs without improvement in site labour productivity.

These factors had different quantitative effects depending on plant location, process technology, project definition, and project management, but an average effect of 25% was reported for UK process industry projects during the 1970s [3]. Somewhat greater effects have been observed from a detailed analysis of cost data for ethylene and chlor-alkali plants [11].

3.5 Data sources for plant and equipment costs

3.5.1 Plant costs

Potential data sources for plant costs include those available privately within a company or on a confidential basis through contractors or licensors, those available on a semi private basis through subscription such as the technology reviews published commercially, and those available in public literature. In the last category, examples include the technological encyclopaedias, reports by OECD, United Nations and national governments on specific technologies, books, and chemical engineering journals. A summary of these sources is provided in Table 3.2.

Table 3.2 Some published sources of plant cost data.

A. Encyclopaedias of Chemical Technology, for example
- Kirk Othmer [12]
- Ullmanns [13]

B. Text books, for example
- Table 6.2. Process cost correlations in Towler and Sinnott [14]
- Table 6.11 Capital cost data for chemical and petroleum processing plants in Peters, Timmerhaus, and West [15]

C. Journals, for example
- Chemical Engineering
- Chemical Engineering Progress
- Hydrocarbon Processing
- Oil and Gas Journal

D. Public statements by investing companies around the time of investment commitment or plant commissioning.

It must be emphasised that while useful as approximate indicators, such data is only a guide. It is often very difficult to assess the reliability or usefulness of published cost data. Adequate definition is required covering:

- location and site characteristics;
- time reference and corresponding business conditions;
- details of raw materials and product qualities and compositions, and temperature and pressure conditions of supply or storage;
- details of process technology employed, plant design, and production capacity;
- composition and mass flowrates of effluents;
- energy efficiency encompassing extent of heat recovery and utilities consumptions;
- inclusions and exclusions, including the distinction between inside and outside battery limits;
- terminal points;
- plot plan and space requirements;
- specific contract terms and conditions which plant owners may impose on contractors for liabilities, quality assurance, and documentation.

In reviewing plant cost data it is important to know whether the data is based on an ***achieved*** or ***estimated*** cost. For an estimated cost, what was the purpose and the basis of the cost? What was the scope of the estimate covering the spectrum of physical aspects of process and utility equipment, piping, instruments, electrical supply and transmission, buildings and structures, and so on. What were the competency and experience levels of the estimators? Rarely in publicly available cost data are all of these important definitive details provided. It then becomes a matter of judgement as to the reliability and usefulness of the data.

Some reported project costs and estimated costs for process plants fulfilling different functions are listed in Table 3.3. These examples are listed to emphasise the dependence of capital costs on process technology employed, feedstock composition, plant location, and the scope of the project. The examples are briefly discussed.

Example A. Treatment of metallurgical smelter gases containing sulphur dioxide

Capital and operating costs have been estimated by Glonka and Brennan [16] for capturing sulphur dioxide from metallurgical smelter gases at various Australian locations to produce sulphuric acid, sulphur or gypsum, as well as for downstream processing of sulphuric acid into phosphate fertilisers. Here, of key importance are the input concentrations of sulphur dioxide, the need for wet gas cleaning to remove other impurities in the gas, and the design of the sulphuric acid plant. Site location for a smelter is also important influencing:

- potential increase in treatment costs and product transport costs to markets for cases where smelting is done at the mine site;
- cost of transporting mineral concentrate to a different location for smelting and downstream processing.

Table 3.3 Some estimated and achieved fixed capital costs for selected projects.

Function	Production plant and feedstock	Production capacity	Location	Fixed capital costs
Treatment of SO_2 in metallurgical smelter gas	Sulphuric acid produced from 12% (v/v) SO_2	362 kt/year H_2SO_4 inclusive of wet gas cleaning	Elura mine site Midwest NSW Australia	$42.8 million IBL $5.3 million OBL $Aust 1992
	Saleable gypsum produced from 3% (v/v) SO_2	734 kt/year $CaSO_4$ inclusive of wet gas cleaning		$32.6 million IBL $4.7 million OBL $Aust 1992
Electricity generation	Black coal Steam turbine	1000 MW	Australian city or industrial site	$2.29 billion $Aust 2000
	Brown coal Steam turbine	1000 MW		$3.18 billion $Aust 2000
	Natural gas Combined cycle	1000 MW		$1.31 billion $Aust 2000
	Natural gas Open cycle	1000 MW		$1.54 billion $Aust 2000
	Black coal IGCC	394 MW		$1.31 billion $Aust 2008
Natural gas extraction, treatment, supply	BassGas From 150 km offshore	20 PJ/year NG, 65 kt/year LPG, 1 million bbl/year condensate	Offshore/onshore pipelines to Victoria	$750 million $Aust 2006
	Thylacine Otway from 70 km offshore	75 PJ/year NG, 100 kt/year LPG, 0.8 million bbl condensate	Offshore/onshore pipelines to Victoria	$810 million $Aust 2004
Petroleum refining	Crude oil distillation Atmos. pressure	100 MBPSD feedstock	US petroleum refinery	US$80 million $US 2005
	Catalytic cracking	80 MBPSD feedstock		US$260 million $US 2005
Ammonia	Steam reforming of NG; incl. electricity, steam generation, NH_3 storage	0.76 million t/year NH_3	Pilbara, Western Australia	$700 million $Aust 2006
	Steam reforming of NG	0.85 million t/year NH_3	Louisiana, USA	US$850 million US$ 2016
Hydrogen	Coal	380 t/day H_2	Australian city or industrial site	$648 million $Aust 2011
	Natural gas	380 t/day H_2	Australian city or industrial site	$219 million $Aust 2011

Example B. Electricity generation
Capital and operating costs have been estimated for electricity generation using brown coal, black coal, and natural gas in steam turbine (ST), combined cycle (CC), and open cycle (OC) systems [17], and for black coal using integrated gasification combined cycle (IGCC) system with and without carbon capture [18]. Capital costs reflect both the fuel used, the generation cycle and efficiency achieved, and as for the IGCC case, the processing of the syngas.

Example C. Natural gas extraction and treatment
Offshore natural gas must be extracted and piped to onshore treatment facilities, involving both offshore and onshore piping. Treatment facilities reflect impurities in wellhead gas (e.g. CO_2, H_2S) as well as the extent of co-product recoveries, and the location of both offshore source and onshore treatment facility. Two examples of projects involving gas piped from offshore Victoria and treated onshore to meet sales gas specification are listed [19].

Example D. Petroleum refining
Capital and operating costs in petroleum refinery units are reviewed by Gary and others [9]. As well as the distinct feedstock and product quality influences there are technology options which influence the various unit processing costs, for example continuous catalyst regeneration or fixed bed units for catalytic reforming units.

Example E. Ammonia production
Capital costs for ammonia production will depend on location, plant design including energy efficiency and project scope, for example the extent of utility generation and ammonia storage. Recent large-scale projects of interest include Incitec Pivot's plant in Louisiana drawing on shale gas and Yara Pilbara fertilisers drawing on offshore natural gas from Western Australia [20, 21].

Example F. Hydrogen production
Location, plant design, energy efficiency, and project scope also influence the capital cost of hydrogen plants. A further influence for both hydrogen and ammonia plants is the choice of feedstock. Capital costs for hydrogen plants using natural gas and coal are compared in Table 3.3. The comparison was part of a wider study of the environmental and economic consequences of natural gas substitution in feedstock and fuel applications, in the context of exploring resource depletion assessment methodology [22].

3.5.2 Equipment cost data

Equipment costs have frequently been presented in chart form. Costs are usually correlated with a capacity parameter, for example the surface area for heat exchangers or filters, or absorbed power for pumps or compressors, using the relationship

$$I = k_i X^b \tag{3.6}$$

where

I = equipment cost
X = capacity parameter
b = exponent
k_i = proportionality factor

This relationship is of the same form as used for complete plants, though the exponent b tends to have a somewhat wider range of values than for plant costs. The relationship is also frequently applied to installed equipment costs. An extensive study of 662 cases found 71% of values of exponent b within the range 0.4–0.8 with the average value of 0.6 [23]. Despite the wide use of the relationship in practice, its reliability may often be limited. As with plant costs, there is a strong tendency with most equipment types for the exponent to increase with larger equipment capacities until some practical or economic limit is reached. There may also be a lower capacity limit below which it is not practical or economic to purchase or fabricate.

Alternative equations for correlation have been used. For example, Breuer and Brennan [24] have adopted a correlation of the form

$$\log(\text{Cost}) = a + b(\log X) + c(\log X)^2 + d(\log X)^3 \tag{3.7}$$

The correlation enables the direct graphical representation of data on logarithmic coordinates, and a quantitative measure of the economy of scale at any value of the capacity parameter through the derivative term $d(\log \text{Cost})/d(\log X)$.

Equipment cost data may be obtained from published sources, accumulated in-house company data, or equipment vendor quotations. Some useful sources of published equipment cost data from chemical engineering texts are summarised in Table 3.4. Most published data are from the United States, though there are some from other countries. The data requires application of location and inflation factors, using a similar approach to that for entire plants. Rather less variation in equipment cost by country may be expected since if significant cost differences occur for various countries, items can generally be imported as required.

Of the textbook sources listed in Table 3.4, Peters et al. [15] provide a detailed and comprehensive database. Costs are presented graphically, offering the advantages of showing change in the cost vs capacity exponent over the size or capacity

Table 3.4 Some textbook sources of equipment cost data.

Couper, Penney, Fair, and Walas [25], 'Process Engineering Economics'.
Towler and Sinnott [14], 'Chemical Engineering Design Principles, Practice and Economics of Plant and Process Design', 5th edition.
Turton, Baille, Whiting, Shaelwitz, and Bhattacharyya [26], 'Analysis, Synthesis and Design of Chemical Processes'.
Peters, Timmerhaus, and West [15], 'Plant Design and Economics for Chemical Engineers', 5th edition.
Breuer and Brennan [24], 'Capital Cost Estimation of Process Equipment'.

range, and the available capacity range for the relevant item. The comprehensive database is also an advantage in demonstrating the options within particular equipment categories for equipment selection and specification.

Imported items require cost adjustments for freight and tariff. Even though a country may have fabrication capability, a purchaser from within that country may prefer to import equipment for reasons of assured quality, even if tariff penalty occurs.

Much cost data are reported for equipment constructed in a single material (usually carbon steel) and for a given design pressure. If other construction materials or design pressures apply, correction factors must be used as multipliers of the base equipment cost. Some indicative correction factors for pressure vessels and heat exchangers are provided in Table 3.5, but these assume both a given set of material costs and a given proportion of materials and labour costs in equipment costs; hence they serve only as a guide. As equipment size increases, for instance, materials cost becomes a more dominant component of equipment cost and the materials correction factor increases. Some literature sources, for example Peters, Timmerhaus, and West, provide a range of costs for different materials and design pressures for a diversity of equipment items.

Figs. 3.2 and 3.3 show some examples of correlated equipment costs for pressure vessels and shell and tube heat exchangers constructed in carbon steel. The data has been taken from two Australian projects implemented in 1988–1989. While the data is historic, it is reproduced because of the extent and scatter of the data points. Scatter is rarely reported in published equipment cost data and is an indication of potential variation in equipment costs actually achieved in projects, and hence in the reliability of estimates. The shapes of the two sets of curves also support the evidence of variations in the relationship between capacity index and equipment size for different equipment types.

Table 3.5 Indicative correction factors for some materials of construction.

Material	Materials factor for process vessels	Materials factor for heat exchangers
Carbon steel	1.0	1.0
Type 304 stainless steel	2.2–2.8	1.5–2.2 CS shell, Type 304 SS tubes 2.8 Type 304 SS shell, Type 304 SS tubes
Type 316 stainless steel	2.9–3.5	1.8–2.5 CS shell, Type 316 SS tubes 2.8–3.2 Type 316 SS shell, Type 316 SS tubes

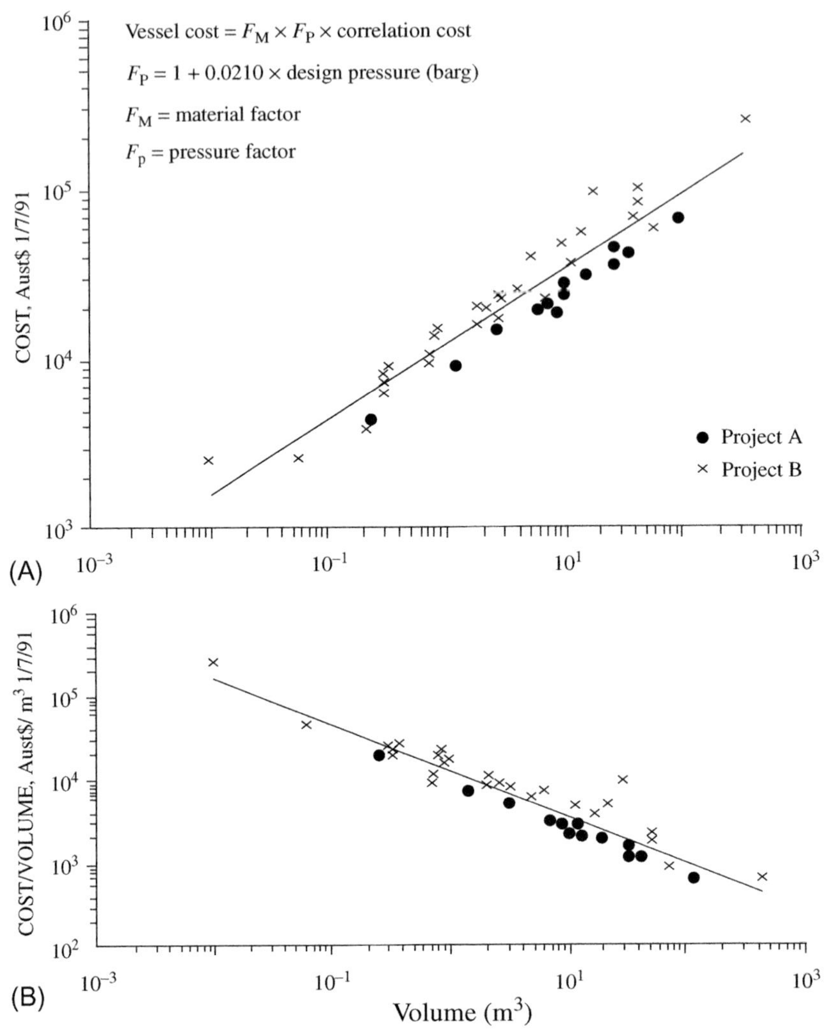

FIG. 3.2

Correlation of (A) cost and (B) cost per unit volume for pressure vessels of carbon steel construction. Cost read from graph requires adjustment for design pressure and materials other than carbon steel.

In addition to engineering influences, the equipment selling price will be affected by business factors. These include the availability and cost of materials and fabrication labour, the required delivery time specified by the purchaser, the extent of loading (or capacity utilisation) of the fabrication shop at the time of fabrication, various purchaser characteristics relating to terms and conditions of purchase, stringencies in design, inspection, and procurement requirements including any imposed liquidates

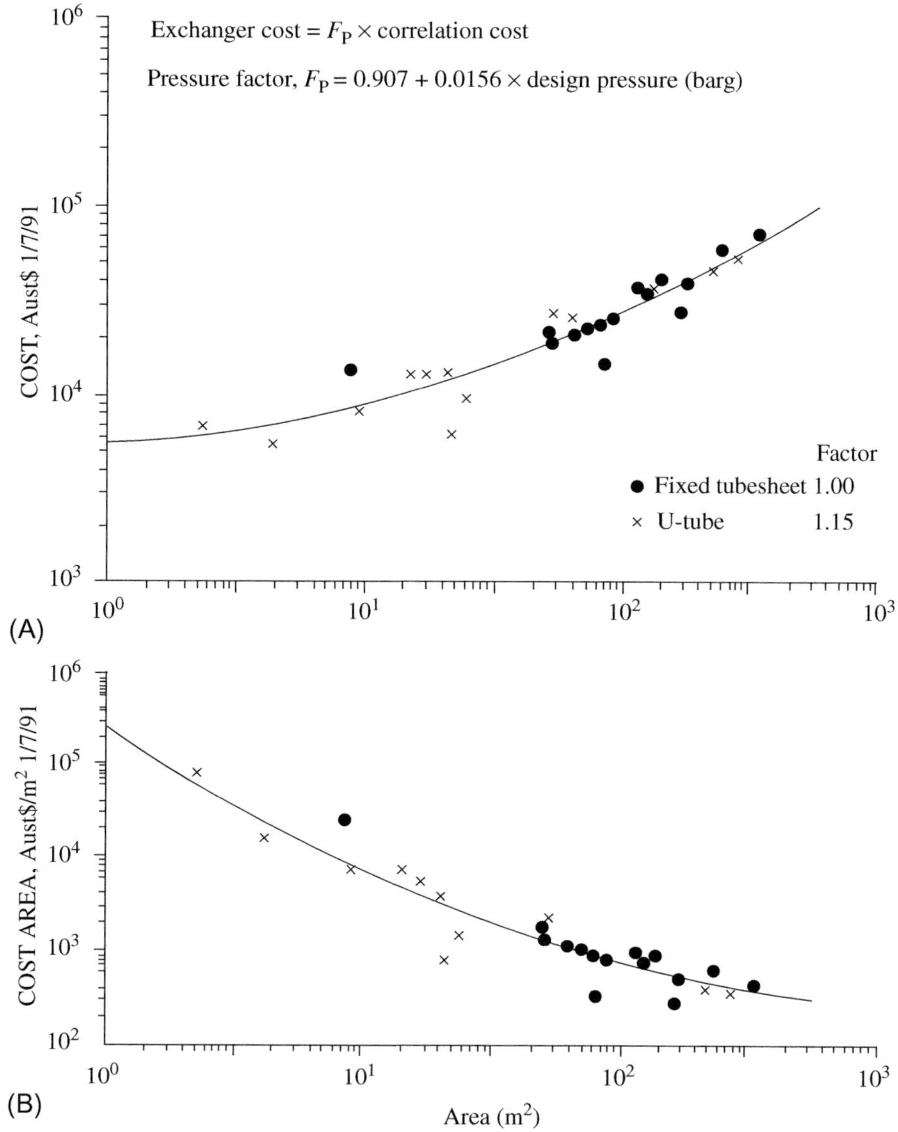

FIG. 3.3

Correlation of (A) cost and (B) cost per unit area for shell and tube heat exchangers of carbon steel construction and 19.05-mm diameter tubes.

damages and risks, established buyer-seller relationships, and multiple purchase item discounts. The effect of such business factors operating collectively with an adverse effect can result in the purchased cost of an equipment item being as much as 100% in excess of a base case situation (Breuer and Brennan [23]).

Rules of thumb are often used for vessels constructed in common materials expressed as dollars per fabricated tonne. Thus for pressure vessels, some common yardsticks are US$9000 per fabricated tonne for mild steel construction and US$18,000 per fabricated tonne for stainless steel construction, as of 2018. These are very approximate guides, not accounting for scale effects (which can be major) or design influences, and more applicable to 'one-off' fabrication. While such yardsticks must be used with caution, they can be useful reference points in early costing or in evaluating quotations.

It is important to adopt a critical approach to published equipment cost data. In most reference book sources, for example the following aspects are not addressed:

- the business conditions pertinent to the cost data;
- the volume of cost data on which the correlation or graphical representation is based;
- the scatter in the data points leading to the correlation or graphical representation.

Further, there is often a lack of engineering definition accompanying the cost data. For example, the published cost data for centrifugal pumps often neglects to define the standard of pump construction and fails to specify pump speed, a key parameter governing volumetric flowrate and developed head. Business-related and engineering details should be defined from the viewpoints both of using data and in compiling it for future use.

Ultimately, at the time of project implementation, accurate cost data will become available from equipment vendors. At the time of purchase, consideration must also be given to the likely reliability of the item in operation, its performance efficiency, its maintenance requirements and life expectancy, availability of spare parts, and the reliability of the vendor.

3.6 Plant cost estimation by factoring purchased equipment costs

This approach requires a defined process flowsheet with equipment items specified in relation to function, size, design pressure and temperature, materials of construction, type of construction, and size and type of equipment drive (if applicable). This detail then enables estimates of equipment costs, to which factors can be applied to build up estimates of plant cost. Costs of drives (usually electric motors or steam turbines) for machinery items should be included as part of the purchased equipment cost. Factors may be applied in several ways, two of which are outlined here.

3.6.1 Factors applied to the sum of purchased equipment costs

In the first approach, factors may be applied to the total purchased equipment cost (PEC) to estimate the IBL process plant cost (PPC) as expressed in Eq. (3.1). The factor may be applied as a sum of the component factors (i.e. as an overall *Lang factor*) or using the component factors contributing to the overall Lang factor considered individually. Values of the component factors may vary considerably for projects of varying size, scope, and process technologies. A common approach in chemical engineering texts has been to multiply the sum of purchase equipment costs by an overall Lang factor suitable for either a fluids processing, solid-fluids processing, or solids processing plants. Two examples are those by Peters et al. [15] and Towler and Sinnott [14] shown in Table 3.6. The overall multiplier here includes both inside and outside battery limits costs. Care must be exercised in the use of such factors obtained from different sources and texts to understand the engineering scope of work covered by the particular factor. Differences in scope definition are used by different authors, and the sum of factors employed in some cases include both inside and outside battery limits costs, while in other cases include inside battery limits costs only.

Table 3.6 Multipliers of purchased equipment cost for total fixed capital investment.

Type of processing plant	Peters et al. [15]	Towler and Sinnott [14]
Fluids	5.93	6.00
Fluids-solids	5.03	6.05
Solids	4.67	4.55

Table 3.7 shows the breakdown of **inside battery costs** for 11 Australian projects spanning the late 1980s to mid-1990s with total capital values between $20 million and $120 million ($Australian 2001). These projects were predominantly from oil and gas processing. Cost elements are expressed as factors of purchased equipment cost. Means and standard deviations for each cost category have been calculated for the 11 plants. The ratio of standard variation to the mean is a measure of variability in each cost element.

The variability can be seen to be lowest for the direct plant cost and overall Lang factor, but considerably higher for many individual contributing costs. With the experience of many projects, judgement can be used in factor selection to improve the accuracy of the estimate.

3.6.2 Factors applied to individual equipment items

In a more detailed approach, factors are applied to each individual equipment item, considered separately, to estimate the installed cost of the item. This approach is outlined in 'A Guide to Capital Cost Estimation' published by the Institution of Chemical Engineers [27]. Each equipment item cost is first adjusted to an equivalent cost based on carbon steel construction. Factors are then applied for erection, piping, instruments, electrical, civil, structures and buildings, and lagging. The factored costs are then added to the purchased equipment cost to give an installed equipment cost. These items are then summed to give a 'direct plant cost'. Overhead costs

Table 3.7 Capital cost distributions for some Australian process industry projects [27].

Cost category	Average relative value	Standard deviation	Variability (standard deviation/ average)
Equipment	1	–	0
Equipment installation	0.19	0.06	0.32
Piping	0.60	0.19	0.31
Instrumentation	0.23	0.11	0.46
Electrical	0.14	0.08	0.58
Civils	0.22	0.1	0.47
Structural steel	0.14	0.09	0.61
Buildings	0.048	0.075	1.58
Insulation/ fireproofing	0.089	0.066	0.74
Painting	0.014	0.011	0.78
Direct plant cost	2.70	0.52	0.19
Engineering and project management	0.25	0.065	0.26
Total IBL cost	3.40	0.68	0.20

Note: Values may not sum precisely to totals shown due to rounding.

covering engineering and project management as well as a contingency allowance are added as percentages of direct plant cost to provide an estimate of total plant cost. Values of individual factors reproduced from the Institution of Chemical Engineers booklet [27] are listed in Table 3.8.

While the factors in Table 3.8 are strictly only applicable to the UK costs, Table 3.8 is useful in several respects:

- It offers guidelines for selecting appropriate factors. For example, piping factors are higher for gas processes than liquid processes, reflecting the lower densities of gases relative to liquids and hence larger pipe diameters. The guidelines provided are by no means exhaustive and can be usefully supplemented by experience and data from previous projects, if available. For example, more precise piping estimates would reflect the number of connecting pipes to a vessel, or the spacing between equipment items. A list of such considerations was proposed by Brennan and Golonka [28] in Table 3.9.
- Table 3.8 accounts for the influence of equipment cost reflecting equipment size on the value of the factor employed. For example, instrumentation for a small diameter column implies a larger factor than for the same instrumentation on a larger diameter column. This is to be expected since the instrumentation cost is largely independent of column size.

Table 3.8 Individual factors of purchased equipment costs for estimation of installed equipment cost (Gerrard [27]).

Currency	**Pounds sterling January 2000**						
Value of individual equipment item as purchased standardised to carbon steel basis	**Over 300,000**	**100,000–300,000**	**40,000–100,000**	**20,000–40,000**	**6000–20,000**	**3000–6000**	**Under 3000**
Purchased equipment item	1	1	1	1	1	1	1
Delivery to site							
Installation of equipment item							
Much of site erection included in purchased cost	0.013	0.03	0.04	0.06	0.075	0.09	0.25
Average erection	0.05	0.08	0.1	0.11	0.13	0.15	0.38
Equipment requires some site fabrication (e.g. large pumps require lining up serpentine coolers)	0.08	0.1	0.13	0.15	0.18	0.2	0.48
Equipment requires much site fabrication (e.g. large distillation columns, furnaces)	0.3	0.38	0.45	0.56	0.67	0.77	1.13
Piping including installation							
Ducting and chutes	0.03	0.05	0.1	0.18	0.28	0.43	0.59
Small bore piping or service piping	0.06	0.13	0.26	0.43	0.69	1.04	1.4
Average bore piping and service piping (e.g. mainly liquid piping)	0.16	0.26	0.4	0.66	0.98	1.4	1.76
Large bore piping and service piping (e.g. mainly gas)	0.2	0.33	0.49	0.78	1.11	1.58	1.94

Continued

Table 3.8 Individual factors of purchased equipment costs for estimation of installed equipment cost (Gerrard [27])—cont'd

Currency	Pounds sterling January 2000						
Value of individual equipment item as purchased standardised to carbon steel basis	**Over 300,000**	**100,000–300,000**	**40,000–100,000**	**20,000–40,000**	**6000–20,000**	**3000–6000**	**Under 3000**
Average bore piping with complex system (e.g. much manifolding and recirculation)	0.2	0.33	0.49	0.78	1.11	1.58	1.94
Large bore piping with complex system	0.25	0.41	0.61	0.96	1.38	1.96	2.43
Multiplying factor for materials	as appropriate for piping material specified						
Steam tracing	1.3	1.3	1.3	1.3	1.3	1.3	1.3
Instrumentation							
Local instruments only	0.03	0.04	0.06	0.13	0.24	0.43	0.75
1 controller and instruments	0.09	0.13	0.22	0.34	0.49	0.65	1
2 controllers and instruments	0.13	0.2	0.33	0.45	0.6	0.79	1.14
3 controllers and instruments	0.18	0.33	0.43	0.6	0.77	0.96	1.38
Electrical							
Lighting only	0.03	0.03	0.03	0.06	0.1	0.13	0.19
Lighting and power for ancillary drives (e.g. conveyors, stirred vessels, air coolers)	0.09	0.1	0.14	0.26	0.34	0.41	0.6
Lighting and power excluding transformers and switchgear for machine main drives (e.g. pumps, compressors, crushers)	0.13	0.18	0.25	0.33	0.43	0.51	0.63

Table 3.8 Individual factors of purchased equipment costs for estimation of installed equipment cost (Gerrard [27])—cont'd

Currency	Pounds sterling January 2000						
Value of individual equipment item as purchased standardised to carbon steel basis	**Over 300,000**	**100,000–300,000**	**40,000–100,000**	**20,000–40,000**	**6000–20,000**	**3000–6000**	**Under 3000**
Lighting and power including transformers and switchgear for machine main drives	0.19	0.25	0.34	0.46	0.6	0.74	1
Civil							
Average civil work, including plant and structure foundations, floors, services	0.08	0.1	0.14	0.17	0.22	0.28	0.35
Above average civil work, complicated machine blocks, special floor protection, elevator pits in floors, considerable services	0.15	0.21	0.31	0.4	0.5	0.6	0.85
Multiplying factor for piling	1.3	1.3	1.3	1.3	1.3	1.3	1.3
Structures and buildings							
Negligible structural work and buildings	0.012	0.025	0.025	0.04	0.05	0.06	0.08
Open air plant at ground level with some pipe bridges and minor buildings	0.06	0.08	0.1	0.14	0.17	0.21	0.26
Open air plant within a structure	0.14	0.24	0.31	0.41	0.5	0.59	0.74
Plant in a simple covered building	0.19	0.29	0.39	0.48	0.56	0.69	0.85

Continued

Table 3.8 Individual factors of purchased equipment costs for estimation of installed equipment cost (Gerrard [27])—cont'd

Currency	Pounds sterling January 2000						
Value of individual equipment item as purchased standardised to carbon steel basis	**Over 300,000**	**100,000–300,000**	**40,000–100,000**	**20,000–40,000**	**6000–20,000**	**3000–6000**	**Under 3000**
Plant in an elaborate building on a major structure within a building	0.35	0.48	0.63	0.76	0.9	1.06	1.38
Lagging							
Lagging for service pipes only	0.012	0.03	0.04	0.06	0.1	0.15	0.23
Average amount of hot lagging on pipes and vessels	0.03	0.04	0.08	0.14	0.21	0.31	0.38
Above average amount of hot lagging on pipes and vessels	0.04	0.06	0.1	0.17	0.26	0.35	0.4
Cold lagging on pipes and vessels	0.06	0.1	0.15	0.25	0.31	0.41	0.56

- The approach accounts for the influence of materials of construction for an equipment item. For example, a carbon steel tower might require similar foundations to a stainless-steel tower of the same dimensions. If the stainless-steel tower were not brought to a carbon steel equivalent for factoring foundation costs, the foundation cost estimate would be excessive.

It is evident from Table 3.8, that the value of the overall Lang factor will depend on the average purchased equipment cost and hence on the size of the plant. Thus one may anticipate much higher overall Lang factors for pilot plants or small demonstration plants where the value could conceivably be as high as 10, than for large-scale plants where the value could be as low as 2.5. The dependence of the Lang factor on average equipment cost implied in Table 3.8 is consistent with reports by other authors (e.g. Montfoort and Meyer [29]).

Table 3.9 Additional considerations in factoring equipment costs (Brennan and Golonka [28]).

Check list for I Chem E method	Other considerations
Installation of equipment item	
Much of site erection included in purchase cost	Equipment weight, height, complexity, fragility
Average erection	Number and complexity of internals (trays, catalyst, etc.)
Equipment requires some site fabrication (e.g. large pumps require lining up serpentine coolers)	Ease of access; plant layout Weather, labour productivity, and cost Greenfield/brownfield site
Equipment requires much site fabrication (e.g. large distillation columns, furnaces)	New plant or modification to existing plant; if existing plant, extent of availability for modification work Extent of shop assembly Extent to which unit is packaged
Piping including installation	
Ducting and chutes	Weight, material, pressure (and temperature) rating
Small bore piping or service piping	Extent of fittings, instrument connections
Average bore piping and service piping (e.g. mainly liquid piping)	Packaged unit? Some piping included? Above ground/underground
Large bore piping and service piping (e.g. mainly gas)	Height of equipment item
Average bore piping with complex system (e.g. much manifolding, recirculation)	Stand-alone/clustered (i.e. shared pipe rack) or addition to existing rack
Large bore piping with complex system	Above ground or underground piping
Multiplying factor for materials	Number of branches on equipment
Steam tracing	Spacing between equipment/plant layout Number of valves; valve design, materials, size, complexity
Instrumentation	
Local instruments only	Packaged unit? Some instruments included?
1 Controller and instruments	Extent of alarms, trips, interlocks
2 Controllers and instruments	Transmission
3 Controllers and instruments	Cost of measuring instrument—analyser, temperature, pressure, flowrate, level Computer hardware and software costs Some materials factors for on stream instruments

Continued

Table 3.9 Additional considerations in factoring equipment costs (Brennan and Golonka [28])—cont'd

Check list for I Chem E method	Other considerations
Electrical	
Lighting only	Ground space, equipment height, security/ safety
Lighting and power for ancillary drives (e.g. conveyors, stirred vessels, air coolers)	Considerations for lighting Number of ancillaries requiring power
Lighting and power excluding transformers and	Number of starter stations
Switchgear for machine main drives (e.g. pumps, compressors, crushers)	Size of electric motors Spacing between equipment/plant layout
Lighting and power including transformers and switchgear for machine main drives	Plant electrical class or zone, e.g. 'explosion proof'
Civil	
Average civil work, including plant and structure	Weight, height of equipment
Foundations, floors, services	Wind loadings
Above average civil work, complicated machine blocks, special floor protection, elevator pits in floors, considerable services	Load bearing properties of soil Large momentum components (large, high speed compressors, centrifuges), reciprocating or vibrating machinery
Multiplying factor for piling	
Structures and buildings	
Negligible structural work and buildings	Opportunity for shared structures between similar items—e.g. columns, exchangers
Open air plant at ground level with some pipe bridges	
And minor buildings	Size, height, complexity of equipment
Open air plant within a structure	Buildings to house noisy equipment
Plant in a simple covered building	Explosion proof buildings, control rooms
Plant in an elaborate building on a major structure	Special provisions for on-site maintenance
Within a building	
Lagging	
Lagging for service pipes only	Insulation for noise
Average amount of hot lagging on pipes and vessels	Process temperatures
Above average amount of hot lagging on pipes and vessels	
Cold lagging on pipes and vessels	

Table 3.9 Additional considerations in factoring equipment costs (Brennan and Golonka [28])—cont'd

Check list for I Chem E method	Other considerations
Painting	
	Corrosion protection (function of plant environment) Surface area of plant. Function of equipment supply
Miscellaneous considerations of general applicability	
	Single equipment items or items in parallel Solid, liquid, or gas processing Safety and environmental standards Operating cost inputs Utilities consumptions: utility related investment into piping (steam and cool. water) electrics (electric power) Prevailing business environment in engineering industry
Home office costs	
Design, engineering, procurement, project management	Previous experience; extent to which design is duplicated or modified from previous project or is completely new Capability, experience of engineering team and its management Productivity, salary, payroll overheads of personnel Quality and extent of project definition, especially at commencement of project Extent of client, consultant involvement vs autonomy

Similar tables could be developed for other countries if supporting data were available. In the absence of supporting data, international cost relativities (in the form of those indicated in Table 3.2) may be used to modify both the equipment cost ranges and the factors; clearly, these adjustments become increasingly risky without the support of a reliable database.

Factored estimates can be usefully supplemented in various components of the estimate by costing rules of thumb for materials and labour. For example, estimates for piping materials ($/t) and piping installation (workhours/t and $/workhour) can be used as a cross-check for factored piping costs; estimates of cost per control loop or per alarm/trip system may likewise be used as a cross-check for instrumentation. This approach is often termed 'benchmarking'.

3.6.3 Relevance of factored cost estimates to chemical engineering

While purchased equipment costs which provide the baseline for factored estimates may account for only 30% of IBL capital costs and possibly 20% of total capital costs for a project, factored estimates based on equipment costs are important in chemical engineering contributions to early project development. This is because the plant capital cost estimates are derived directly from equipment specifications, which in turn are derived from a well-defined process flowsheet. Accompanied by further design detail derived from piping and instrumentation diagrams and plant layout drawings, judgement can be made about the various factors which are applied to estimate the total IBL costs for a process plant or for the installed cost of one or more specific equipment items, as indicated in Fig. 3.4.

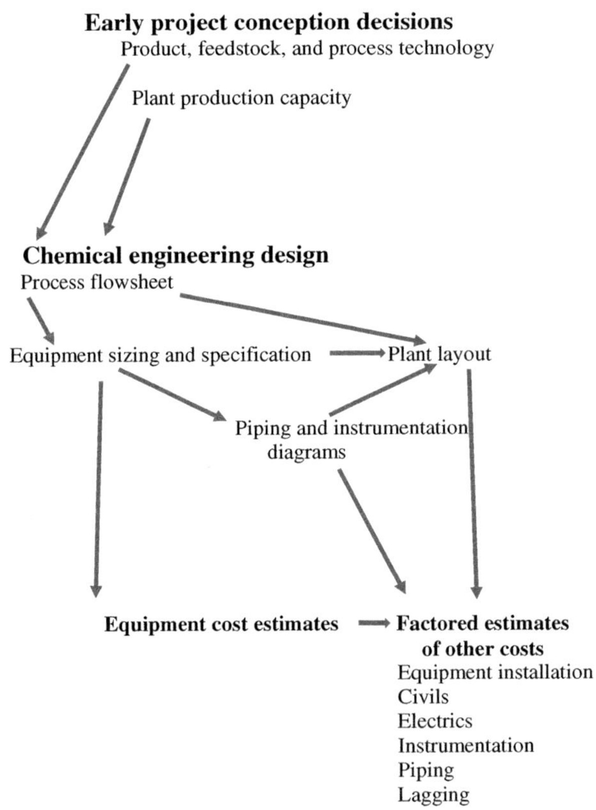

FIG. 3.4

Information flow diagram linking early project and design decisions with initial capital cost estimates.

More accurate estimates demand expertise beyond that of the chemical engineer, relying, for example on expertise and professional experience of:

- specialist engineers in relation to the scope of work required;
- electrical engineers for the cost of electrics;
- civil engineers for the cost of foundations and structures;
- mechanical engineers for piping costs;
- instrumentation specialists (in collaboration with chemical engineers) for instrumentation and control costs;
- construction personnel for construction costs;
- cost estimators, planners, and schedulers for cost rates and project duration estimates.

Dedicated estimating engineers with the benefit of training and experience have a key role in drawing on such expertise and on accumulated data from previously completed projects.

3.7 Detailed estimation of plant and equipment costs

More detailed methods of estimating plant costs, such as those used by a contractor tendering for a job or in controlling expenditure on a job in progress, require considerable design detail. Following a thorough and complete process design, piping, instrumentation, and electrical specifications will be developed, assisted by a plant layout. Civil engineering design will be undertaken including foundations, structures and buildings, assisted by a detailed evaluation of the site. Site construction will be evaluated including labour rates, costs of accommodation if necessary, and temporary site facilities for construction personnel. A detailed plan of design, procurement, construction, and commissioning activities will be developed.

Costs are commonly broken down into materials and labour components, and estimated quantities of materials detailed. Quotations from subcontractors for items such as piping, electrics, instrumentation, and main plant construction will be sought. Past records are invaluable in guiding and checking estimates, and use is made of rules of thumb for checking quantities and labour hours. Parametric-type estimating, which is intermediate between factored and detailed estimating, is often used as a faster approach to supplement detailed estimating; for example if the length of piping is known, data from previous projects in \$/m can be used to estimate piping costs, inclusive of costs of valves and fittings.

For the estimation of fabricated costs for vessels, towers or heat exchangers, a detailed estimate requires detailed drawings and specifications giving materials, dimensions (including thicknesses), branches, internals, and construction details. Materials quantities and fabrication workhours can be estimated, and unit costs are incorporated to estimate a fabricated cost. Allowances for overheads and profit margins need to be added to estimate the purchased cost of the item.

3.8 Computer applications in fixed capital cost estimation

A number of packages have been developed both for in-house and commercial use. Some of the commercial packages are adjuncts to process simulation packages such as ASPEN.

3.9 Cost of revamps

The revamping of an existing operating plant may be expected to have a different cost structure to that of a new plant on either a greenfield or brownfield site. Revamps may have a number of purposes:

- accommodate changes in feedstock, for example a refinery adapting to a heavier, sourer crude oil source;
- accommodate changes in product quality or spectrum of products produced;
- incorporate technological change offering cost savings or other benefits;
- change product packaging or transport modes;
- improve the safety or environmental performance of the plant;
- increase the production capacity of the plant in response to market growth.

Because there are the imposed constraints of an existing plant and site, the engineering design and project management costs of a revamp may be much higher as a proportion of plant costs than for a new plant. While much of the existing equipment may be reused without modification, a number of items typically require modification; vessels may need new internals, reactors may need new catalysts, columns may require new trays or packings, compressors or pumps may need to be increased in speed, and so on. Such equipment modifications save on new equipment costs but incur some component costs and high engineering and construction costs.

Many revamp projects incorporate major improvements in process control leading to high instrumentation costs relative to the purchased cost of new equipment. Modifications to plant layout may incur increased piping costs, leading to high factors for piping. Since plant downtime is often a critical consideration, a need to use intensive labour on a continuous (24 h/day) basis for modifications and construction will incur additional costs, particularly where downtime scheduling is complicated or unpredictable.

Since the objectives and the constraints for revamps are so diverse, cost structures may be expected to be far more variable than for new plants, and conventional factors used in factored estimates become correspondingly less reliable. Detailed knowledge of the existing plant and its site are important inputs to realistic cost estimating for plant revamps.

3.10 Estimation of start-up capital

Start-up costs include all nonrecurring costs between completion of plant construction and commencement of satisfactory operation (with regard to production capacity, product quality, and overall process efficiency). These costs may be apportioned between capital costs and operating expenses but include:

- materials, utilities, labour, and overhead costs in checking, purging, testing and proving of plant;
- process operator training;
- additional costs incurred in modifying plant to facilitate its operability.

Total start-up costs have been reported by a number of published and unpublished sources to be within the range of 1%–10% of fixed capital investment of the plant. The magnitude of the costs are difficult to estimate with precision and some contingency should be provided. Higher costs may be anticipated where the technology is newly developed or where there is a lack of experience with the type of plant concerned.

3.11 Estimation of working capital

Working capital may be defined as funds over and above fixed capital, start-up capital and land investment needed to start and maintain plant operation. Working capital includes inventories of raw materials, finished products and materials in process, the balance between accounts receivable (also called debtors) and accounts payable (also called creditors), and cash.

An accountant's definition of working capital is net current assets, or current assets minus current liabilities. Current assets comprise inventories, trade debtors, cash, and assets a company intends to convert into cash in its course of business. Current liabilities comprise trade creditors and short-term borrowings. Current assets and current liabilities are identified in a company's balance sheet, published as part of the financial statement in a company's annual report. Table 3.10 is an example of such a balance sheet [30] summarising the company's assets, liabilities, and equity. Note the following aspects of the balance sheet:

- the identification of current assets and current liabilities;
- the supplementary notes provided on inventories;
- the contribution of current assets and current liabilities to total net assets;
- the balance between total net assets and total shareholders' equity.

Such balance sheets from company reports are a useful source of information on working capital requirements for process industry companies. Some important aspects of working capital are as follows:

- Although usually smaller in magnitude than fixed capital investment (typically 10%–20%), the initial investment into working capital is spent in a short space of time, at (or immediately before) plant start-up.
- Characteristically, working capital can be readily converted into cash and is normally assumed recoverable at the end of project life without loss. This implies a lesser degree of risk than for fixed capital, which may give rise to a separate financing arrangement than that for fixed capital.
- As production increases, working capital requirements tend to increase, occurring as an incremental change in cash flow. The increase results from an increase in accounts receivable relative to accounts payable, and frequently from an increase in stocks.

Table 3.10 Example of a balance sheet for a chemical company.

		1988 $million
Current assets	Cash	47.9
	Receivables	478.9
	Investments	21.9
	Inventories[a]	593.9
	Other	31.7
	Total current assets	**1174.3**
Noncurrent assets	Receivables	12.2
	Investments	9.9
	Property, plant, and equipment	902.5
	Intangibles	77.1
	Other	134.3
	Total noncurrent assets	**1136.0**
Total assets		**2310.3**
Current liabilities	Creditors and borrowings	610.4
	Provisions	304.7
	Total current liabilities	**915.1**
Noncurrent liabilities	Creditors and borrowings	181.9
	Provisions	133.9
	Other	40.6
	Total noncurrent liabilities	**356.4**
Total liabilities		**1271.5**
Net assets		**1038.8**
Shareholders' equity	Share capital	291.3
	Reserves	348.5
	Retained profits	280.9
	Shareholders' equity attributable to ICI Aust	920.7
	Minority shareholders' interest in subsidiaries	118.1
	Total shareholders' equity	**1038.8**
[a]Inventory contributions		
Raw materials and stores		**192.9**
Work in progress, at cost		**16.4**
Finished goods		**384.6**
Total inventories		**593.9**

Based on ICI Australia, Annual Report, 1988.
Balance sheet as at September 1988 with simplifications.

While working capital is sometimes expressed as a percentage of fixed capital, it is more directly related to operational considerations. Frequently it is expressed as a percentage of annual sales revenue or annual cash operating cost. Working capital

Iw may be estimated from the following relationship, where the components are in currency units.

$$\mathrm{Iw} = \mathrm{Srm} + \mathrm{Sfp} + \mathrm{Smip} + \mathrm{Ar} - \mathrm{Ap} \tag{3.8}$$

where

Srm = raw material stocks
Sfp = finished product stocks
Smip = materials in progress inventories
Ar = accounts receivable
Ap = accounts payable

We now explore the contributions of each of these elements.

3.11.1 Raw material stocks

Raw material stocks are valued at the purchased cost (accounting for delivery). The required quantity of stock will depend on the source of raw material, its mode of delivery, the reliability of supplies (availability pattern for a plant), and the process technology. Even for plants producing the same product, there can be variations due to feedstock differences. For example, an ethylene plant based on ethane feedstock supplied by pipeline will require less raw material inventory than a plant based on naphtha feedstock shipped from a remote refinery. Similarly, chlor-alkali plants using naturally occurring brine pumped from underground require less raw material inventory than plants using salt shipped from remote salt fields.

Broad guidelines by Scott [31] recommend:

(i) for bulk commodities supplied daily or by pipeline—1.5 weeks usage*
(ii) intermittent delivery—3 weeks usage.
(iii) delivery by ship—6 weeks usage.

**Note*. This implies some storage at the plant in conjunction with pipeline supply. In many cases, especially with gas pipelines, pipelines deliver direct to the plant with no additional storage. In these cases, the inventory is limited to that in the pipeline and can be considered negligible in cost terms.

3.11.2 Finished product stocks

Finished product stocks include stocks at the production plant, but if considering the total sphere of operation, stocks in the warehousing and distribution chain as well. Some repackaging may be an integral part of the distribution. The mode of product despatch and the operating needs of customers will influence stock requirement.

Broad guidelines by Scott [31] recommend:

(i) for products with simple specifications delivered daily or by pipeline—1.5 weeks usage
(ii) for products in multiple grades, or with complicated specification—3 weeks storage

(iii) for batch processes in campaigns, where campaigns are lengthened to enable economic utilisation of plant—6 weeks storage.

Extra storage is required where products are sold overseas. The value of a product is assessed at cash costs, that is, operating cost excluding depreciation, representing its contribution to capital investment.

3.11.3 Materials in progress inventory

Materials (or work) in progress inventory varies depending on equipment (or process) residence times, the extent of recycle, the number of reaction steps, and the need to characterise (that is, determine the properties of) intermediate products. The value of materials in progress will be intermediate between raw materials and finished product values. Rules of thumb are more difficult in this case for both stock quantities and values. In the absence of definitive data, an assumption of 1–2 weeks of annual production with a material value taken as the average of raw material and product cash cost is a guide.

3.11.4 Accounts receivable (or debtors)

If finished products are assumed to be paid for at the end of the month following delivery, average indebtedness period is 6 weeks. For exports, the period could be longer, say up to 12 weeks. The value of each product is taken at its selling price.

3.11.5 Accounts payable (or creditors)

If materials or utilities are assumed paid for at end of the month following delivery, the average period of credit is 6 weeks. Wages and salaries make a smaller contribution being paid say on average, 1.5 weeks in arrears.

3.11.6 Other contributions and considerations

Wages and salaries, stocks of maintenance materials and packaging materials, cash in hand, liabilities for tax payment, and employee benefits have been omitted from this analysis. Some of these items may need consideration in special cases but can generally be ignored in a preliminary estimate for a project.

In estimating working capital for a number of plants comprising a process route, where those plants are integrated on the one site, the product stocks for plant A become the raw material stocks for Plant B. Thus distinction must be made between site integrated and nonsite integrated manufacture.

Special consideration is necessary for plants producing co-products (e.g. chlorine/caustic soda, phenol/acetone) in estimating product stock value. An initial approach might be to proportion production cost relative to tonnages produced, although where one product involves more processing than another, appropriate adjustment should be made.

Working capital estimates are usefully detailed on a worksheet. An example of a blank worksheet is provided in Fig. 3.5. This needs to be used in conjunction with the

Working capital estimating worksheet				
Product				
Annual production, t/yr				
Raw material stocks				
Raw material	Inventory, weeks	Unit usage, t/t	Unit cost, $/t	Value, $million
Total				
Materials in process				
Inventory				
Material	Inventory, weeks production	Inventory, t	Unit value, $/t	Value, $million
Total				
Product stocks				
Product	Inventory, weeks		Product cash cost, $/t	Value, $million
Total				
Debtors	Period, weeks		Product selling price, $/t	Value, $million
Creditors	Period, weeks	Cost, $/t product		Value, $million
Raw materials				
Utilities				
Subtotal				
Total working capital				

FIG. 3.5

Example of a working capital cost estimating sheet.

operating cost worksheet discussed in Chapter 4. Worked examples involving working capital considerations and estimates are provided in Chapter 6.

References

[1] AACE International, Cost Estimate Classification System, Recommended Practice No. 18R-97, November 29, 2011.
[2] D.H. Allen, R.C. Page, Revised technique for predesign cost estimating, Chem. Eng. 82 (1975) 142–150.
[3] C.A. Miller, Current concepts in capital cost forecasting, Chem. Eng. Prog. 69 (5) (1973) 77–83.
[4] J.H. Taylor, P.J. Craven, Economic evaluation at the predesign stage, I. Chem. E. Symposium Series No. 55, 1978, pp. 3/1–3/19.
[5] A.M. Al-Fadli, M.A. Soliman, K.M. Wagialla, I.S. Al-Mutaz, Optimal resources allocation and processing for the Saudi petrochemical industry, Process Econ. Int. 7 (1988) 22–29.
[6] D.J. Brennan, Process Industry Economics. An Australian Perspective, Longman Cheshire, 1990.
[7] G. Martin, Construction costs in South East Asia, HTI Q. (Spring) (1995).
[8] Compass International Inc., Front End Conceptual Estimating Yearbook, 18th ed., 2019.
[9] J.H. Gary, G.E. Handwerk, M.J. Kaiser, Petroleum Refining. Technology and Economics, fifth ed., CRC Press, 2007.
[10] A.T.R. Kjar, Improving the competitiveness of capital intensive industries, in: Institution of Engineers (Aust.) Conference on Engineering Management, Adelaide, 1985, pp. 101–107.
[11] D.J. Brennan, An alternative approach to modelling performance improvement in the process industries, in: Chemical Engineering in Australia, vol. ChE12(4), December 1987, pp. 7–12.
[12] Kirk Othmer Encyclopaedia of Chemical Technology.
[13] Ullmanns Encyclopaedia of Industrial Chemistry.
[14] G. Towler, R. Sinnott, Chemical Engineering Design; Principles, Practice and Economics of Plant and Process Design, second ed., Elsevier, 2013.
[15] M.S. Peters, K.D. Timmerhaus, R.E. West, Plant Design and Economics for Chemical Engineers, fifth ed., McGraw Hill, 2003.
[16] D. Brennan, K. Golonka, Costs and environmental impacts in pollutant treatment: a case study of sulphur dioxide emissions from metallurgical smelters, Trans. IChemE 75 (Part B) (1997) 232–244.
[17] J.R. May, D.J. Brennan, Sustainability assessment of Australian electricity generation, Trans. IChemE 84 (Part B) (2006) 131–142.
[18] T.J. Falke, A.F.A. Hoadley, D.J. Brennan, S.E. Sinclair, The sustainability of clean coal technology: IGCC with/without CCS, Process Saf. Environ. Prot. 89 (2011) 41–52.
[19] S. Rimos, Resource Depletion Impact Assessment by Consequence Analysis—Impacts of a Natural Gas Scarcity in Australia (Ph.D. Thesis), Monash University, 2015.
[20] Yara Pilbara Fertilisers, https://www.yara.com.au/about-yara/about-yara-australia/pilbara/yara-pilbara-fertilisers/.
[21] https://www.incitecpivot.com.au/about-us/major-projects/ammonia-plant-usa.

[22] S. Rimos, A.F.A. Hoadley, D.J. Brennan, Determining the economic consequences of natural gas substitution, Energy Convers. Manage. 85 (2014) 709–717.
[23] J. Haldi, D. Whitcomb, Economies of scale in industrial plants, J. Polit. Econ. 75 (1967) 373–385.
[24] P.L. Breuer, D.J. Brennan, Capital Cost Estimation of Process Equipment, Institution of Engineers Australia, 1994.
[25] J.R. Couper, W.R. Penney, J.R. Fair, S.M. Walas, Chemical Process Equipment and Design, Elsevier/Butterworth-Heinemann, 2012.
[26] R. Turton, R. Baille, W. Whiting, J. Shaelwitz, D. Bhattacharyya, Analysis, Synthesis and Design of Chemical Processes, fourth ed., Prentice Hall, 2009.
[27] A.M. Gerrard (Ed.), Guide to Capital Cost Estimating, fourth ed., Institution of Chemical Engineers, 2000.
[28] D. Brennan, K. Golonka, New factors for capital cost estimation in evolving process designs, Trans. IChemE 80 (Part A) (2002) 579–586.
[29] A.G. Montfoort, F.A. Meijer, Improved Lang factor approach to capital cost estimating, Process Econ. Int. iv (1) (1983) 20–21.
[30] ICI Australia, Annual Report, 1988.
[31] R. Scott, Working capital and its estimation for project evaluation, Eng. Process Econ. 3 (1978) 105–114.

CHAPTER

Operating cost estimation

4

Resources do not excuse us from resourcefulness.
Professor Owen Potter, Emeritus Professor of Chemical Engineering, Monash University, Melbourne, Australia

Chapter outline

Process Industry Economics. https://doi.org/10.1016/B978-0-12-819466-9.00004-3

4.1 Scope and classification of operating costs

In the process industries, operating costs encompass all costs associated with the production, distribution, and marketing of products, together with the ongoing costs of developing or purchasing the enabling technology. Operating costs also include all management and business costs incurred indirectly in making and selling products. Operating costs extend over the operating life of a plant, but are commonly evaluated for a stipulated time period; in project evaluation, the time period is conventionally taken as one year. In reality, operating costs will vary from one year to the next, resulting from any changes in production rates, maintenance schedules or operating routines, or from any plant modifications. Costs will also vary with change in business conditions affecting the prices of raw materials, utilities or labour. Many of these changes cannot be foreseen at the evaluation phase of a project and are often ignored in initial estimates.

During plant operation, operating cost targets must be set and compared with achieved performance as part of the plant management strategy; shorter periods, typically of 3 months duration, are often adopted for budgeting. Operating costs, as well as being expressed in currency units per time period, are widely expressed in terms of currency units per tonne (or similar unit) of product produced.

In addition to defining the boundaries of fixed capital costs, it is important to define the boundaries of operating costs. Considering the battery limits of plant investment, for example, it costs a certain amount to take raw materials delivered to the plant boundary and process these into a product delivered at another point at the plant boundary. Further costs will be incurred in transferring the product to customers. The customer may be an adjacent plant and product transfer may occur through a pipeline. Alternatively, there may be a complex chain of product storage at the manufacturing site, transport from the site which may entail road, rail or shipping or a combination of these, and further storage at intermediate points, before ultimate delivery to the customer. Similarly, raw materials supplied to the process will potentially have undergone a sequence of upstream processing and transport steps, incurring corresponding costs and influencing their unit cost.

In some cases, by-products or effluents generated in the plant will need further treatment, often in a treatment facility located on the same site or in a separate plant. It is most important to define the associated boundaries for manufacturing cost and to account for all product and effluent streams, and their relevant production or treatment costs.

It is necessary in the context of a total business operation to account for all operating costs incurred in manufacture. An accounting distinction is often made between

manufacturing costs which are largely under the control of the plant manager and ***nonmanufacturing costs***. Nonmanufacturing costs include the costs of product distribution, product selling, ongoing research and development, and that share of the cost of running the wider company which can be allocated to the product or business in question.

Manufacturing costs encompass a list of identifiable costs, but these may be broadly classified as:

- raw materials
- utilities
- personnel employed for the diverse range of tasks required
- capital-dependent costs such as maintenance, insurance, and property taxes

These costs may be conveniently grouped in a simplified model for production costs expressed in Eq. (4.1). This model is useful as a means for making a quick, approximate estimate of manufacturing cost, in identifying dominant cost contributions, and in assisting an understanding of the ways in which operating costs are classified. Since utility costs are energy-intensive, it may sometimes be convenient to substitute energy consumption for utilities consumption in the model.

$$C = \sum R_i r_i + \sum E_j g_j + (Mm)/QU + (kI)/QU \tag{4.1}$$

where

C = production cost (\$/t product)
R = consumption of raw material i (t raw material/t product)
r = unit cost of raw material i (\$/t raw material)
E = consumption of utility j (e.g. MWh electricity/t product)
g = unit cost of utility j (e.g. \$/MWh)
M = number of employees/t product
m = average annual cost per employee (\$/person) including payroll overheads
k = factor to account for a number of costs dependent on fixed capital
I = fixed capital investment (\$)
P = annual production (t product)
Q = annual production capacity (t product)
U = capacity utilisation (P/Q)

From the model, QU is the annual production achieved, Mm/QU represents the personnel costs per tonne of product, I/QU the fixed capital investment per tonne of product produced annually, and kI/QU the fixed capital-dependent costs per tonne of product.

It is constructive to distinguish between ***performance parameters*** R, E, M/QU as distinct from ***unit cost parameters*** r, g, m. Performance parameters depend on the technology adopted and on plant management, whilst unit cost parameters depend largely on influences outside the control of the company operating the plant, and are often location dependent. In discussing changes in the four cost categories in Eq. (4.1) contributing to production cost, it is important to analyse whether the as arisen because of a change in performance or in unit cost, or changes in ese. Fixed capital investment per unit of capacity is a function both of per- (technology and management) and unit costs (labour and materials). The l-dependent costs are also functions of both performance and unit costs. on (and all operating costs) are classified according to their dependence on rate. For a plant having a given design (or rated) production capacity, those costs expressed in \$/year which vary with change in production rate are described as ***variable***, whilst those which are unaltered with the change in production rate are described as ***fixed***. In this context, we need to express our simplified model of Eq. (4.1) in terms of costs per time interval (or dollars per year) as

$$\underset{\text{Production cost}}{C(QU)} = \underset{\text{Raw materials costs}}{\sum (R_i r_i)(QU)} + \underset{\text{Utilities costs}}{\sum (E_j g_j)(QU)} + \underset{\text{Personnel costs}}{Mm} + \underset{\text{Capital-dependent costs}}{kI} \tag{4.2}$$

For a continuous process plant, raw materials and energy costs are classified as variable, and personnel- and capital-dependent costs as fixed. In preliminary analyses, raw materials and energy costs tend to be represented as linearly variable, and personnel- and capital-dependent costs as completely unchanged, with variation in production rate. Detailed analyses, however, may show these simplified models to be inaccurate. For example:

- raw material and energy productivities ($1/R$ and $1/E$) may well decline as production rate increases, especially when production rate approaches or even exceeds design capacity
- electricity consumption in MWh per tonne of the product may be much greater at low production rates than the linear model predicts, due to the need to keep machinery operating and sustain cooling water reticulation, lighting, or refrigeration at low production rates
- wear of construction materials (and hence maintenance costs) may increase on some plants with increasing production rates
- packaging labour requirements may increase at higher production rates

Nevertheless, the concept of fixed and variable costs is broadly useful and is frequently expressed diagrammatically in conjunction with sales, to show the dependence of profit on production rate (see Fig. 4.1). This highlights the importance of operating at a production rate close to production capacity, that is, at a high level of capacity utilisation, U.

Capacity utilisation, the ratio of production rate achieved to production capacity is an important consideration in operating cost and profitability evaluation. As production rate decreases to progressively lower levels of capacity utilisation, the plant eventually operates at a loss; if this situation perpetuates, the plant may be shut down because of its poor profitability. Capacity utilisation may be restricted by several influences such as:

- lack of demand (the market influence)
- poor plant reliability

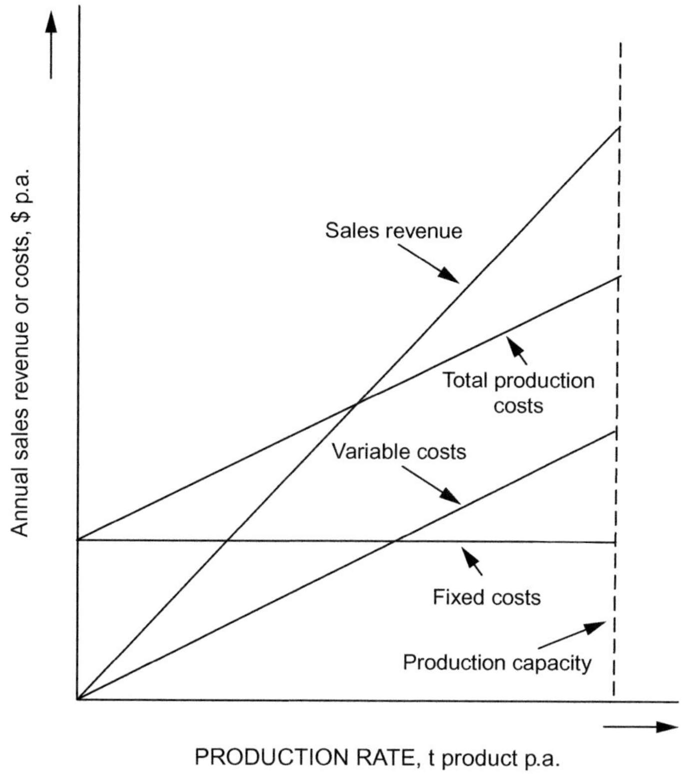

FIG. 4.1

Dependence of sales revenue and production costs on production rate.

- shortages of raw materials or
- labour productivity problems, including the effects of strikes

Raw materials availability is a particular issue in the evaluation of processes based on renewable feedstocks, such as bagasse (used as a feedstock for bioethanol) where seasonal influences affect the pattern of supply. Shipping frequency can be another issue where feedstocks are imported and stored in large quantities. Seasonal influences on product demand can also affect capacity utilisation, for example, in the supply of natural gas, where demand for domestic and commercial heating varies from summer to winter.

A further cost classification widely used in the context of varying production levels is that of ***marginal*** or ***incremental cost***, which is the additional cost per tonne of product resulting from an incremental increase in production. The marginal cost per tonne of product is usually much less than the total cost per tonne of product at the increased production level, and in many continuous process plants, may be limited to the variable costs of production.

As well as being dependent on production rate, production costs are scale (or capacity) dependent, reflecting the dependence of both fixed capital investment and operating personnel requirements on plant capacity. Thus whilst R and E are generally independent of plant capacity, M/Q and I/Q generally decrease with increasing capacity, providing ***economies of scale***. The extent of dependence of production cost on capacity depends both on the relationships between I and M with Q, and the magnitude of personnel and fixed capital-dependent costs compared with raw material and utility costs.

It is a common practice amongst chemical engineers to summarise operating costs as an operating cost summary sheet, which identifies all the key contributors and their inputs in terms of performance parameters and unit cost parameters, with costs estimated on an annual basis and as cost per tonne of product. An example of such a cost summary sheet for preliminary estimates is shown in Table 4.1. The spectrum of cost categories contributing to total operating costs and the basis and means of their estimation are now explored.

4.2 Raw material costs, consumptions, and prices

Raw materials are materials which are fed into the process at various stages and are ultimately converted by physical and chemical (and sometimes biochemical) changes to products (and in some cases by-products). Major raw materials are readily identifiable, but minor raw materials must also be accounted for. Thus salt is a major raw material in the manufacture of chlorine and caustic soda, but other chemicals are also required, for example, soda ash for reducing calcium concentration in brine fed to chlorine cells and sulphuric acid for drying wet chlorine produced in the chlorine cells. The term '***feedstock***' is frequently used for a major starting raw material in a process, especially in the context of hydrocarbon processing.

At the evaluation phase of a project, raw materials consumption may be estimated from flowsheet considerations based on detailed mass and energy balances. It is important however to account also for operating inefficiencies derived from factors such as the production of out-of-specification product, and a range of potential production losses. Potential losses may be due to evaporation, spillage, release through purge streams, or weather effects in outside storage.

Such inefficiencies are difficult to quantify without the benefit of operating experience. At the operational phase of a project, overall mass and material balances can be compared with achieved inputs and outputs; this is an important aspect of both process and environmental management, since leaks, spillages, and fugitive emissions contribute both to materials losses and environmental impacts.

Data on raw materials consumption as well as other performance parameters for established commercial processes are available in process technology reviews. Such

Table 4.1 Simplified operating cost summary sheet.

Product				
Process route				
Plant capacity				
Capacity utilisation				
Product selling price				
Fixed capital				
Production cost	Unit usage (unit/t)	Unit cost ($/unit)	**Annual cost ($million)**	**Cost per tonne product ($/t)**
Raw materials				
....				
....				
Total raw materials cost			....	....
Utilities				
Electricity				
Fuel				
Steam				
Cooling water				
Other				
Total utilities cost			....	....
Process labour	**Number**	**Salary ($/year)**		
Operators per shift				
Number of shift teams				
Total shift operators				
Day operators				
Total process labour wages			....	....
	% Process labour wages			
Payroll overheads			....	....
Plant overheads			....	....
	% Fixed capital			
Maintenance			....	....
Insurance			....	....
Property taxes			....	....
Book depreciation			....	....
Total fixed cost			....	....
Total production cost (TPC)		Fixed	....	....
		Variable	....	....
		Total	....	....
	% TPC			
Corporate administration	....		....	....
Research and development	....		....	....
Selling expenses	....		....	....
Total operating cost			....	....

data always needs careful assessment in the light of the scope and technology adopted in the relevant process plant. Process technology reviews can be found in

- encyclopaedias of chemical technology (e.g. Kirk Othmer [1] and Ullmanns [2])
- books (e.g. McKetta [3], Rudd et al. [4], and Gary et al. [5]) as well as
- journals related to chemical engineering and the process industries.

Some indicative raw materials requirements for selected process technologies are listed in Table 4.2. Some indicative price data for raw materials spanning fuel and mineral commodities as well as some intermediate chemicals are provided in Tables 2.4–2.6 in Chapter 2. Some important considerations in assessing price data are discussed in Sections 2.5.2 and 2.5.3; important strategic drivers from a project design viewpoint include the source, quality, ease of access, and supply reliability of the raw material.

Table 4.2 Indicative raw materials requirements for selected process technologies.

Product	Raw material	Consumption Tonnes per tonne product
Acetic acid	Methanol	0.53
	Carbon monoxide	0.47
Alumina	Bauxite	2.0
	Caustic soda	0.1
	Lime	0.05
Aluminium	Alumina	1.93
	Anode carbon	0.5
Ammonia	Methane	0.57
Chlorine and caustic soda	Salt	1.68
Diammonium phosphate	Ammonia	0.22
	Phosphoric acid	0.65
Ethylene	Ethane	1.24
Ethylene	Propane	2.11
Ethylene	Naphtha	2.7
Ethylene oxide	Ethylene	0.88
	Oxygen	1.1
Formaldehyde	Methanol	1.18
Hydrogen	Methane	3.2
Methanol	Methane	0.71
Nitric acid	Ammonia	0.28
	Oxygen	0.17
Phosphoric acid	Phosphate rock	2.43
	Sulphuric acid	2.0

Table 4.2 Indicative raw materials requirements for selected process technologies—cont'd

Product	Raw material	Consumption Tonnes per tonne product
Single superphosphate	Phosphate rock	0.63
	Sulphuric acid	0.39
Soda ash	Salt	1.5
	Limestone	1.24
	Coke	0.09
Sodium cyanide	Methane	0.35
	Ammonia	0.37
	Sodium hydroxide	0.85
Sulphuric acid	Sulphur	0.32
Vinyl acetate	Acetic acid	0.71
	Ethylene	0.35
Vinyl chloride monomer	Chlorine	0.64
	Ethylene	0.49

4.3 Utilities

Utilities (sometimes referred to as services) include primarily energy, heating and cooling needs for the process, but also inert gas and in some cases effluent treatment facilities. The more common utilities in a process plant are fuel, steam, electricity, cooling water, raw water, compressed air, nitrogen, and refrigerated coolants.

4.3.1 Utilities consumption

At the evaluation phase of a project, **steam** and **cooling water** consumption may be estimated from flowsheet considerations, based on detailed mass and energy balances. Allowance for losses and contingencies must be made. In the case of steam, energy losses due to radiation and line losses are probable and may account for up to 20% of theoretical calculations, whilst for batch processes careful accounting must be made for sensible heat requirements. Steam may be required in certain cases for turbine drives on machinery. For cooling water, losses and contingencies may also account for approximately 20% of theoretical calculations. Cooling water is often required for a range of duties associated with machinery components, for example, lubricating oil coolers on gas compressors, which are not accounted for in flowsheet energy balances.

Electricity consumption for a process plant may be estimated from the sum of electric power consumptions on the drives for each operating equipment item. At design production capacity, this generally approximates 80%–90% of the installed power for operating drives. Additional allowances are required for line losses and contingencies, which collectively may account for 10%–15% of that estimated for equipment drives. Thus total power frequently approximates the sum of rated capacities of operating drives (excluding capacities for standby equipment). The rated power is multiplied by the anticipated online hours of the plant for the year, to give an annual energy consumption in kilowatt hours at design annual production capacity.

Electricity consumption varies considerably for process plants. For electrochemical processes such as chlorine and caustic soda from sodium chloride, and for aluminium smelters converting alumina to aluminium, high consumptions of electricity are required for reactors and are major contributors to operating costs. Electricity consumption is also high for high-pressure processes such as ammonia production, due to the need for synthesis gas compression.

Raw water is required as make-up for steam circuits (where demineralised water is normally used) to replace steam losses and boiler blowdown, and also for cooling water circuits to replace evaporation, entrainment, and purge losses. Water is also required for washing, drinking, and sanitary purposes. It may be drawn from one of a number of sources, for example, mains, river, bore, or other. Desalinated sea water is increasingly being used for remote mining or processing sites, and for supplementing infrequent supplies of rain water and low levels in storage dams [6].

Fuel (typically natural gas, coal, or petroleum-derived products) may be required for furnaces, kilns, and driers, as well as for steam raising. Fuel is often required when high temperatures are required beyond those achievable from superheated steam. Knowledge of equipment design will be necessary to assess heat losses and to estimate fuel consumption. Fuel may also be required for start-up of certain plants; for example, where a catalytic reactor operates at elevated temperatures, preheating of the catalyst and reactor is required.

Purge gas and **compressed air** requirements vary considerably for different plants. Some **refrigerant** losses may occur in refrigeration plants, especially at start-up and early operation, needing make-up of refrigerants. Estimates of purge gas, compressed air, and refrigerant consumptions need to be guided by previous plant operating experience.

4.3.2 Unit costs of utilities

A number of approaches can be adopted in obtaining the supply of utilities. Utilities may be purchased from a public or private utility, from an adjacent plant on the same site (e.g. steam raised from process heat recovery), from a company-owned centralised facility (e.g. a steam and power plant providing steam and electricity to a number of different plants on the one site), or generated on site solely for one process plant.

In each case, the utility cost structure can be quite different in relation to contributions of fuel, water, labour and capital costs, and profit margin. For example:

- **raw water cost** will depend on availability, quality, treatment and supply costs
- **desalinated sea water cost** for raw water usage will depend on the scale of the desalination plant and energy source

- **cost of cooling water derived from freshwater** will comprise predominantly make-up water, energy for cooling tower fans and cooling water circulation pumps, and capital-dependent costs (see Example 6.2 for problem on cooling water cost)
- **cost of sea water cooling** will have a different cost structure to that derived from freshwater; limits on return water temperatures to the ocean and resulting pumping costs, and corrosion impacts on heat exchanger capital costs must be considered
- **steam costs** will reflect the costs of fuel, make-up demineralised water, capital-dependent costs derived from the boiler, chemicals used in boiler feedwater treatment, energy consumed in pumping water to boilers, and labour. Steam costs will reflect the pressure of steam generation

The simplified model for production costs represented in Eq. (4.1), applies also to the cost of utilities production. The cost of **electric power** will depend on the technologies employed in its generation and the cost of distribution. Hydroelectric power has been traditionally the cheapest. Costs from fuel combustion systems are influenced by the particular fuel used, the capacity and technology of generation, capacity utilisation, the cost of distribution and supply, and in some countries by imposed carbon taxes. Renewable sources such as solar are being increasingly explored for integration within a spectrum of mining, fuel extraction, and process industries.

Table 4.3 summarises some published capital and operating costs, value-added, and CO_2 emissions for fuel and generation cycle systems used in Australia. The estimates indicated that natural gas combined cycle was most competitive in both capital and operating costs and CO_2 emissions. These estimates were based on a gas price of \$4(Aust.)/GJ which has since doubled in parts of Australia since the estimates were made arising from LNG export commitments. The brown coal steam turbine system

Table 4.3 Estimates of Australian electricity generation costs [7].

Generation cycle-fuel system	Steam turbine Brown coal	Steam turbine Black coal	Open cycle Gas turbine Natural gas	Combined cycle Gas turbine Natural gas
Capital investment (\$billion)	3.18	2.29	1.54	1.31
Value added (\$/MWh)	57.3	42.7	35.9	45.8
Annualised cost (\$/MWh)	46.4	66.9	50.4	34.7
CO_2 emissions (kg/MWh)	1209	1023	825	535

Notes: (1) Cost estimates are in \$Aust. 2000 and based on 1000MW generating capacity. (2) Value added = sales value less nonlabour and noncapital inputs. (3) Annualised cost = annualised capital cost + operating and maintenance costs.

was observed to be competitive economically in terms of value-added because of the low cost of brown coal, but has major capital requirements and is the largest emitter of CO_2.

In some cases, utility prices charged to consumers involve both fixed and variable cost components, reflecting the cost structure of generation and distribution. This approach is commonly adopted in domestic supply of electricity and water and may also be applicable to industrial consumers. In other cases, utilities may be charged as a simple unit cost (e.g. $/GJ, $/MWh, $/t). Since both fixed and variable costs are incurred in utilities generation and supply, scale economies are realised. The extent of the scale economies reflects the cost structure for the particular utility.

Table 4.4 provides some indicative current costs of utilities for industrial use. These are indicative only and will vary for different countries and plant locations. The costs reflect the availability, quality and cost of fuel and water, as well as capital costs of generation facilities and labour costs. Prices can also be influenced by quantities purchased and approaches to accounting for capital charges. In some cases, public utilities include the cost of distribution in their price structure, so that there is no price benefit from being close to the source of utility generation (nor penalty from being remote). Public utilities often negotiate special contracts for large users. Electricity tariffs usually include concessions for users of off-peak electricity. In some cases, electricity prices are influenced by an agreed maximum demand; if the upper limit is exceeded, price penalties are incurred.

Table 4.4 Indicative prices of utilities supplied to process plants.

Utility	Cost US$ 2018	Comments
Natural gas	$3–8/GJ	Price is cheapest in United States (approximately $3/GJ) reflecting availability and price of shale gas; European prices are higher (approximately $5–6/GJ); prices in Japan are higher again (approximately $9/GJ) reflecting cost of liquefaction at exporting country, shipping and regasification
Steaming coal	$60–85/t	
Steam		
HP—40 bara	$10–16/t	
MP—15 bara	$8–13/t	
LP—5 bara	$6–9/t	
Electricity	$70–140/MWh	
Water		
Towns water	$0.7–1.8/t	Water costs vary considerably. Costs of desalinated water are reported to have decreased in recent years, i.e. from 2015 to 2019
Desalinated water	$1.0–4.0/t	
Boiler feedwater	$1.5–2/t	
Cooling water	$0.1–0.2/t	
Nitrogen	$30–70/t	

The European Commission [8] has published a series of reports on energy prices and costs in Europe, comparing prices both in European member countries and countries globally. The report also explores the driving forces for costs, such as the generation technology employed, fuel costs, approaches to energy efficiency, and strategies for mitigating carbon emissions.

4.4 Personnel costs

Fig. 4.2 indicates the variety of personnel functions required for the operation of a process plant. These functions could be grouped as process labour, maintenance labour, and a diverse range of staff.

Process labour includes process operators (frequently on shift) responsible for plant operation including start-up and shut-down routines as well as abnormal operation. It is important to note that the total number of shift process operators will be four to five times the number of operators per shift to enable a viable shift roster. Additional process labour over and above shift labour may be required on day work for making up reagents, product packaging, or raw materials handling.

Maintenance labour includes tradespersons and their assistants responsible for ongoing maintenance (planned and unplanned) on the plant. Some shift cover may be appropriate in special cases, but most maintenance work is more commonly carried out during week working days, with call-ins to handle abnormal situations if required.

Direct ***supervision of process labour*** and ***maintenance labour*** are normally provided by personnel who have had direct experience of the work involved. Shift supervision is normally required for process labour and day supervision for maintenance labour.

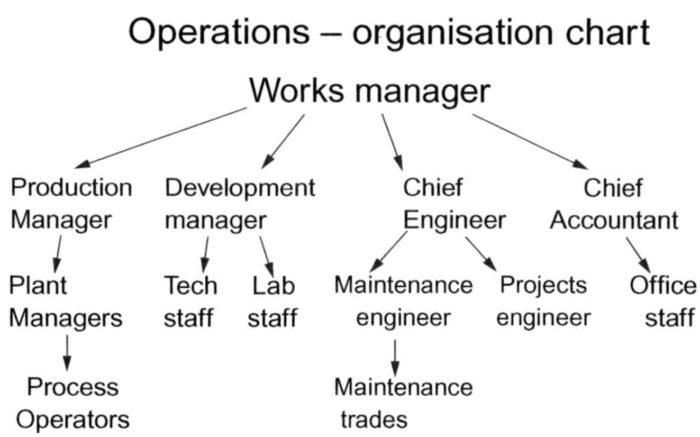

FIG. 4.2

Typical personnel functions required for an operating process plant.

Contract labour is frequently employed for maintenance at major shutdowns, which may be planned on an annual, biennial, or triennial basis depending on the particular plant and its technology. In some cases, contract labour may also be employed for utility plant and process plant maintenance.

Staffing may be broadly categorised as technical, administrative, and clerical. In most process plants this includes production, engineering, technical support, laboratory, accounting, clerical and secretarial functions, as well as specialist functions. Some managerial and specialist functions (e.g. personnel, medical, stores, safety, transport) may be shared between more than one plant.

Some traditional approaches to preliminary estimation of operating costs on process plants have estimated process labour costs initially, and then applied factors for other personnel or personnel-related costs. Since total costs related to personnel have typically accounted for some three times the cost of process labour, a poor estimate of process labour costs can have a considerable influence on the validity of such an estimate.

Plant personnel frequently interact with other company personnel in the areas of marketing, technology, research and development, product sales, and corporate planning. Plant personnel need specialist services to fulfil their responsibilities, and also are required to give specialist advice to other company personnel, as well as outside bodies in industry, community and government, by virtue of their specialised knowledge of the plant.

Divisions between process labour and maintenance labour responsibilities in process plants were once rigidly defined. Some trends in more advanced economies, however, have been towards less rigid demarcation of responsibilities with certain personnel being required to perform both operational and maintenance functions.

In recent years, in established economies a general policy has been adopted to reduce the number of employees in all facets of company organisations. This has been true in process industry companies for all classes of personnel. Particular care must, therefore, be adopted in projecting historic data for plant personnel requirements for use in current estimates.

The influence on personnel requirements from plant and equipment capacity, technology evolution, degree of automation and computerisation, country and site location, and industrial relations policy can be considerable. Insight into these influences can be gained from detailed studies of particular technologies from a global perspective over an extended time period. One such published study has been made of aluminium smelting technology [9].

From the 1950s to the early 1980s, total personnel requirements on aluminium smelters, chlorine/caustic soda plants and ethylene plants employing best available technology and scale did not change greatly despite considerable growth in plant capacity [10]. Chemical industry statistics for OECD over the same period show little growth in overall employment concurrent with substantial growth in scale and diversity of the industry; a slight decrease in the proportion of operating to nonoperating personnel over the period is evident.

Useful information about personnel requirements for a process plant or industry at a given time can often be obtained from published literature, for example environmental impact statements or technology reviews. Some broad observations are now presented.

4.4.1 Effects of process technology, complexity, and scope

As the number of processing steps and complexity increases, so will the demands on operating labour, maintenance labour, and diversity of technical and administrative personnel increase. Apart from the inherent requirements of a particular process technology, drivers for increased personnel numbers include

- extent of raw material purification required
- extent of heat integration and materials recycling
- the number of products produced
- the extent of product packaging
- the extent of effluent treatment
- the extent of integration with other plants on the same site
- the nature and extent of storage facilities for raw materials, intermediates, and products

4.4.2 Effects of capacity

Personnel needs depend on plant capacity. The effects are frequently different for different categories of personnel, as illustrated by some planning studies for aluminium smelters [11, 12]. Aluminium smelters are of interest since they typify parallel streamed processes. These forecasts indicated that where n is the number of potlines and M the number of persons employed:

$M \propto n^{0.4}$ for professional/technical/supervisory staff
$M \propto n^{0.6}$ for qualified tradesmen
$M \propto n^{0.85}$ for process workers
$M \propto n^{0.4}$ for clerical/service staff
$M \propto n^{0.67}$ for total personnel

The number of potlines may be taken as a direct indicator of smelter capacity. At the two potline stage, total personnel comprised 50% process workers, 23% qualified tradespersons, 16% professional/technical/supervisory staff, and 11% clerical/service staff.

Other examples of parallel streamed plants include

- chlor-alkali reactors for converting sodium chloride brine into chlorine, caustic soda, and hydrogen
- the use of membrane cells in sea water desalination plants

In parallel streamed plants, there is a greater dependence of personnel on capacity than in a predominantly single stream plant such as vinyl chloride monomer, or a plant of mixed single and parallel streaming such as ethylene (parallel reactors but essentially single stream purification). Table 4.5 lists some published capacity exponents for various categories of labour in plants using specific technologies; whilst the data is drawn from earlier sources, it illustrates the variation in dependence on production capacity for different employee categories, as well as the high dependence on the capacity for process labour in parallel streamed plants.

As a broad generalisation, the dependence of total personnel employed (M) on plant capacity (Q) approximates the relationship

$$M = k_M Q^a \tag{4.3}$$

where k_M is a proportionality factor

$a = 0.3–0.4$ for single stream plants

0.6–0.7 for parallel stream plants

and has an intermediate value for mixed single stream/parallel stream structured plants.

Table 4.5 Some published capacity exponents for various categories of personnel.

Technology	Category of labour	Capacity exponent	References
Aluminium smelting	Process workers	0.85	Comalco Ltd [11]
	Qualified tradesmen	0.6	
	Professional/technical/ supervisory	0.4	
	Clerical/service	0.4	
	Total manning	0.67	
Aluminium smelting	Works management and production	0.62	Alcoa Portland [12]
	Administration	0.35	
	Engineering	0.2	
	Smelter	0.97	
	Anode plant	0.95	
	Total manning	0.62	
Ethylene	Operating labour	0.1	Wynn and Rutherford [13]
	Maintenance labour	0.7	
	Others	0.7	
	Total manning	0.35	
	Operating plus maintenance labour	0.35	Pilbara [14]
Chlorine/ caustic soda	Process labour and cell maintenance	0.47	Guglielmi [15]
	Operating plus maintenance labour	0.42	Pilbara [14]

As in the case of fixed capital costs, there is evidence in some cases of an increase in the exponent as plant capacity increases.

4.4.3 Effects of time and technological progress

If personnel data is obtained from a previous project, it is important to consider the effect of both the scopes of the plants considered and technological progress in applying the data to a current project. Studies of performance on plants of best available technology and scale over extended periods have shown personnel productivity as the area of most rapid productivity change. Annual rates of change in personnel productivity during periods of plant capacity development approximated 5%, 8%, 9%, and 12%, respectively, for aluminium smelting, chlorine, ethylene, and catalytic cracking plants. Refer to Section 7.8 and Table 7.4 in Chapter 7 for further details. Plant and equipment capacity development and automation are key factors.

4.4.4 Effect of location

Personnel levels on identical plants may be different in various countries and locations due to differing policies and traditions in industrial relations, work practices, and wage rates. In aluminium smelters for example major differences in personnel numbers have been reported in United States or France and developing countries in 1960s; differences are also evident within the 1980s between newly constructed United States and Australian smelters.

4.5 Estimating process labour requirements and costs

Estimating process labour requirements for a process plant require a detailed consideration of operational tasks, including both routine and abnormal operation. This requires experience and knowledge of a particular type of plant and its operation. Reference to known personnel levels for a plant involving similar unit operations is helpful. For example, ammonium nitrate plants involve a neutralisation reaction, followed by evaporation, crystallisation, centrifuging, drying, and packaging. Personnel levels for such plants could be useful in estimating requirements for sodium phosphate or sodium cyanide plants which involve similar unit operations. Gary et al. [5] have provided estimates of operational staff for a naphtha desulfuriser within a petroleum refinery, which could guide estimates for other units in a refinery. Clews has provided details of shift operators for chlor-alkali plants [16].

Wages for process operators vary with the particular type of industry, its technical sophistication, level of profitability and company policy, as well as with the particular country. For many countries, the shift operator's wage has been marginally higher than a chemical engineering graduate's commencing salary. A shift operator's wages are higher than a day operator's wages, reflecting the requirements imposed on working days and times. A comparison of labour rates between countries for 2012 has been published by the US Bureau of Labour Statistics [17]. More recent labour comparison data is published by the US founded Conference Board [18].

4.5.1 Payroll overheads

In addition to wages or salaries paid to employees, other costs are incurred by the employer through workers compensation premiums (related to government safety provisions), leave (including sickness, annual, statutory holidays, and long service), payroll tax, and superannuation pension contributions. These costs are typically 30% to 50% of wages or salaries depending on the employing company and relevant government policy. They also vary from country to country.

4.5.2 Consumable stores

These items comprise gloves and other protective clothing for operators and supervisory personnel, lubricants, and miscellaneous items used in the course of duty. Also referred to as operating supplies, they represent a minor operating cost and are frequently estimated as a function of process labour costs.

4.6 Plant (or works) overheads

These costs include the costs of factory or works staff including engineers, scientists, accountants, clerical, secretarial personnel, specialist personnel, and senior works management. They can also include a range of other costs associated with maintaining stores, vehicles, medical, cafeteria, and workshop facilities, and site security; these costs include outside contract staff and some capital-related costs. The costs of employing factory staff can be estimated from numbers of persons employed and their salaries, with an allowance for payroll overheads. Estimates of staff numbers can be made by extrapolation of data on similar operating plants. Companies sometimes prefer to engage external consultants or contractors to perform tasks which might otherwise be done by employed staff.

Plant overheads are less directly identifiable with process operations than are raw material, utilities, process, and maintenance labour costs. Plant overheads have often been estimated as a factor of process labour, or process labour plus maintenance labour, or occasionally as a percentage of fixed capital in preliminary estimates. The costs associated with laboratory control (laboratory staff, analytical equipment maintenance, reagents) are sometimes excluded from plant overheads and treated as a separate cost category. Wherever such factored approaches are adopted, it is essential for more accurate estimates to consider the total number and roles of employees and their salaries, as a basis.

4.7 Depreciation costs

The concept of depreciation as a cost derives from the loss in value of an asset, in this case, the process plant, over its operating life. The fixed capital cost of the plant is a measure of its initial value. At the conclusion of its operating life, plant value is usually negligible. There may be some salvage value for certain equipment items, but this will be less than the purchased value and less again than the installed value of the equipment. Plant dismantling and/or demolition must be accounted for, and there are some cases (e.g. nuclear power plants) where the dismantling and decontamination costs are substantial. Site decontamination costs for chemical plants and mining operations have recently received increased attention; in many cases, such costs reflect long periods of operation and limited awareness of environmental effects at the time of operation.

Duration of plant life is difficult to estimate with precision at the investment proposal stage. Usually, economic factors, rather than physical deterioration, determine the end of plant life. Product or process obsolescence, lack of competitiveness in scale or cost structure, costs of meeting changes in environmental and safety standards, or excessive maintenance costs are some of the contributing factors to plant obsolescence and decisions by companies to terminate plant operation. Estimates of project life need to take into account the likely product and technology life cycles, the dynamics of the economy, the extent of competition, and physical wear (and corrosion). Physical wear may lead to excessive maintenance costs or poor reliability of plant; there may also be increased vulnerability to safety and environmental risks. For process plants producing undifferentiated products, 10–15 years operating life is a common assumption in investment appraisals. Many process plants, of course, have much longer lives, though expansion and modernisation programmes involving additional capital investment are frequently involved. Power generation plants are assumed to have longer lives (say 20 years) and the Victorian State Government in Australia assumed a 30-year life for the Portland aluminium smelter [19]; both of these projections have been exceeded.

Whilst depreciation can justifiably be regarded as an operating cost, it is important to recognise that depreciation is not a direct component of cash flow, or not an 'out of pocket' expense. It is more strictly an internal company provision to reflect the reduction in the value of plant and equipment over time, and the need to provide funds for future reinvestment. Such reinvestment would be in a replacement plant of improved technology necessary to ensure continuing business competitiveness, or alternatively in another business venture judged to be of greater long-term benefit to the company and its shareholders.

It is important to distinguish between *'book' depreciation* and *'tax' depreciation*. The former is an internal company allowance. The latter is the allowance the tax commissioner will permit the company to deduct from profit to arrive at assessable income for taxation. The tax depreciation allowance thus directly influences the after-tax cash flow, through influencing the tax paid. The two allowances are similar in concept but different in execution and may be quite different in value. 'Book' depreciation affects the company's retention of earnings for future reinvestment, whilst 'tax' depreciation affects the after-tax cash flow and achieved profitability for the project involved.

There are two basic methods for estimating depreciation:

(i) straight-line method
(ii) reducing balance method

The straight-line method depreciates a plant by a constant percentage of the initial fixed capital investment. Thus an investment of \$200 million depreciated over 10 years would have an annual depreciation allowance of \$20 million until the plant was fully depreciated. More generally (assuming zero scrap value)

$$\text{Annual depreciation allowance} = (\text{Fixed capital investment})/n \tag{4.4}$$

where n = economic life in years
and rate of depreciation 'r' is given by

$$r = 100/n \tag{4.5}$$

The reducing balance method employs a constant rate depreciation of the net depreciated value of the asset. It leads to higher annual allowances in the early years. For example, if the constant rate were 15% per annum, an investment of \$200 million would have depreciation allowances of \$30 million, \$25.5 million, and \$21.6 million in the first 3 years. The economic life n can be related to the depreciated rate r by the expression

$$r = 1 - (\text{Residual value}/\text{Intial investment})^{1/n} \tag{4.6}$$

Tax depreciation is discussed further in Chapter 5 and in particular Section 5.2.

4.8 Land taxes

These taxes include land and water rates, which are functions of unimproved and improved capital values and hence the location. These usually comprise a relatively small cost item. Most companies operating process plants find it cheaper and more satisfactory to own land than to lease industrial premises.

4.9 Insurance

Insurance of the plant against fire, explosion or storm damage can be negotiated. The premium is a function of the risk involved and usually negotiated as a percentage of fixed capital. Contributing factors to the risk are complex but include

- the inherent safety of the process concerned
- the design provisions of the plant
- the location of the plant, including its vulnerability to extreme weather events
- maintenance standards
- aspects of management including safety training and operating procedures

Typical percentages range between 0.1% and 3.0%, averaging 0.5% [20]. A detailed account of the principles governing economics and insurance is given by Mannan [21].

Should plant damage be extensive, the elapsed time between damage and recommissioning of the repaired plant will inevitably result in loss of sales revenue.

To cover this loss, an additional annual premium may be paid, termed '***loss of profits insurance***' which usually amounts to 1%–2% of annual sales revenue.

4.10 Maintenance

Maintenance costs include the cost of all materials and labour for planned and unplanned maintenance of plant and buildings. Allowance needs to incorporate major shut downs or turn arounds, for which additional contract labour may be required over and above the regular maintenance employees.

Annual maintenance costs are usually estimated as a percentage of fixed capital. The percentage is higher for processes which operate under more severe conditions of temperature, pressure, and corrosion. Values quoted from various sources range from 2% to 11% fixed capital, averaging around 5%–6%. Increasing emphasis on plant reliability is leading to increased capital costs of plant with correspondingly reduced maintenance costs as a proportion of fixed capital. The breakdown in maintenance costs is approximately evenly divided between materials and labour in most cases. As is the case for personnel cost estimation generally, it is important to cross-check the percentage of fixed capital estimate with an estimate of the likely number of maintenance workers required together with their salaries.

4.11 Royalties

Royalties are payments for technology which is incorporated in the plant design or aspects of the process operation. These may be made as an initial lump sum, or as a continuing royalty based on sales revenue, or a combination of the two. Royalty payments may entitle the user to access technology improvements which can be incorporated into existing or subsequent plants. Royalties vary considerably with the nature of the industry and product, reflecting both novelty and research intensiveness. Royalty payments in some cases may continue until a patent expires.

Licensors of process technology often make agreements with process plant contractors, for the mutual benefit of each party in marketing the process technology, using their respective skills.

4.12 Packaging

Packaging costs include the cost of packages (e.g. drums, bags, cylinders, bottles), ancillary materials (e.g. pallets, shrink-wrapping), packaging labour, and costs associated with maintenance and depreciation of packaging equipment.

Packaging costs may vary considerably in magnitude and in relation to the product value. For some products and plants, packaging is not required. It is therefore impossible to generalise about packaging as a factor of production costs. Little published cost data are available in this area.

4.13 Nonmanufacturing costs

4.13.1 Distribution costs

The distribution cost of a product is usually paid by the consumer, but may be charged to the product cost as a marketing strategy. Distribution costs include costs of transport, all associated product handling including loading and unloading facilities, and the costs of storage at distribution centres. Distribution costs are a function of the product (e.g. a liquefied gas, or free-flowing solid), the scale of transportation, hazards associated with transport, distribution of customers and their relative requirements, the basic methods of transportation (rail, road, shipping, pipeline), and the transportation distance involved.

The costs of distribution become an increased proportion of total operating costs for low-value products, for example, sulphuric acid, and in such cases often provide a form of protection for local producers against import competition.

Little is published in the public domain about costs of transportation and distribution of process industry products. Indications of costs can however often be obtained from a specific industry or product studies which include consideration of transport costs. Costs are often reported in terms of $/t km although in some cases transport costs can be dominated by loading and unloading costs. A transport cost model would need to account for the capital investment in the transportation facility inclusive of storages at loading and unloading points, energy costs incurred in transport, associated personnel costs and insurance costs. There are many contributing influences to actual charges such as the viability of backloading, government subsidies, effect of competition, and the approach to the depreciation of the transport ship or vehicle.

4.13.2 Selling expenses

The costs of marketing and selling products are functions of the nature of the product and the market structure. Related aspects have been discussed in Chapter 2. An intermediate like sulphuric acid or ethylene will require very little product development or sales support compared with an end product like polypropylene. Selling expenses will increase with the number of customers and the diversity of product applications. A relatively new product is likely to require a considerable amount of marketing and technical service both to establish product applications and to encourage the adoption of the product by potential users. Selling expenses typically contribute between 1% and 20% of manufacturing costs. Estimates should be cross-checked against the envisaged personnel numbers and their salaries necessary to provide the service.

4.13.3 Research and development

Costs associated with research and development (R&D) are dependent on the company, its size, ownership structure and policy, and on the types of product, process and technology that characterise the company's business. Companies tend to centralise their R&D activities usually in the country of their incorporation and may undertake little or no R&D in overseas countries in which they have operations. This has been true, for example, of many US-owned petroleum refining and chemical companies which have had overseas operations in smaller countries.

For industries undergoing rapid technological change (e.g. materials development) or involving products with relatively short life cycles (e.g. pharmaceuticals) or industries seeking to embark on higher risk new technologies (e.g. biotechnology), R&D costs are higher. If companies do not elect to do R&D, they must be prepared to purchase their technology through royalties or operate in businesses where technological change is slow.

R&D costs incorporate laboratory and pilot-scale plant costs but are also heavily dependent on highly trained research and development personnel. Most literature sources suggest R&D costs amount to 1%–7% of manufacturing costs averaging around 3%.

4.13.4 Corporate administration

Process plants are owned and financed by companies which need a number of personnel to direct and advise the company in areas such as management, public relations, finance, and corporate planning. These costs typically account for 3%–6% of manufacturing costs.

4.14 Capital recovery and financial expenses

The time value of money and cost of capital is normally accounted for in cash flow estimates using the net present value criterion, as discussed in Chapter 5. However, it is sometimes convenient in operating cost estimation, as an alternative to book depreciation, to include the burden of capital cost using a capital recovery factor. The capital recovery factor accounts for the annual cost of recovering a capital investment over a plant operating life of n years, where the interest cost of capital (on a real basis and after tax) is i%/annum. The capital recovery factor f is applied as a factor of capital investment to give an annual cost and is defined as

$$f = \frac{i(1+i/100)^n}{(1+i/100)^n - 1} \tag{4.7}$$

Thus for a value of i of 8%/annum and a life n of 15 years, $f = 0.117$.

Financial expenses for a project vary depending on the sourcing of the finance, and on the extent to which loan and equity funds are used. Generally, projects are evaluated, at least initially, independently of the financing policy.

4.15 Factored approach to operating costs

Factored approaches to operating cost estimation are intended for quick, approximate estimates, and for comparison of processing alternatives. These approaches have in common the factoring of process labour and fixed capital investment costs for addition to the costs of raw materials, utilities, and (if applicable) product packaging. Manufacturing costs can be broken down into

$$C = C_R + C_U + (1 + k_L)L + k_I I + [P] \tag{4.8}$$

where

C = manufacturing costs
C_R = raw material costs
C_U = utilities costs
L = process labour costs
I = fixed capital expenditure
P = packaging costs, where applicable
k_L = factor of process labour costs, accounting for supervision, payroll overheads, operating supplies, laboratory, and plant overheads
k_I = factor of fixed capital expenditure accounting for depreciation, maintenance, insurance, and property rates

Nonmanufacturing costs can be estimated as a factor k_C of manufacturing costs, such that total operating costs O can be expressed as

$$O = C(1 + k_C) \tag{4.9}$$

where k_C accounts for R&D, selling expenses, and corporate administration.

The factors k_L, k_I, and k_C are, respectively, the sums of smaller component factors which when multiplied by L, I, or C represent various cost categories contributing to the total operating cost. Typical values of overall factors k_L, k_I, and k_C, are 2.8, 0.15, and 0.2, respectively. Whilst these values will differ for plants of different process technologies, their variation is likely to be less than the variation in factors within the contributing costs.

Table 4.6 reflects the author's estimating experience supported by the accumulated wisdom of the literature. Explanatory notes are included in the table to assist in the selection of appropriate factors.

Table 4.6 Operating cost factors.

Operating cost scope	Factor value	Influences on factor value
Factors of labour costs, *L*		
Payroll overheads Work safety premiums, leave provisions, payroll tax, superannuation contributions	$0.3–0.5 \times L$	Depends on company and government policy. Tends to be higher in more developed countries
Consumable stores	$0.05–0.2 \times L$	Higher for more complex, dirtier processes
Supervision	$0.1–0.3 \times L$	Depends on organisational structure

Table 4.6 Operating cost factors—cont'd

Operating cost scope	Factor value	Influences on factor value
Plant overheads Costs of employing accountants, engineers, clerical, administrative and management staff. Costs of consulting services; medical, cafeteria, facilities	0.5–1.5 × L	Depends on organisational structure, plant capacity, company policy, technical complexity, extent of plant or product development, single or multiple products
Laboratory Sometimes a subset of plant overheads includes laboratory personnel and operating cost of analytical equipment and reagents	0.05–0.25 × L	Higher for specialty chemicals, multiple products, processes with complex chemistry
Factors of fixed capital, *I*		
Maintenance Includes cost of materials M and labour Lm. Contract labour may be used for shutdowns. Costs are often similar for materials and labour	0.02–0.12 × I	Higher for corrosive, abrasive, severe duty processes. Higher for plants pushed to capacity limits
Insurance	0.001–0.03 × I	Higher for greater risk
Property taxes	0.01–0.04 × I	Higher for more developed site locations
Book depreciation (not a cash cost)	0.05–0.2 × I	Depends on perceived operating life of plant
Capital loan charges	0.05–0.2 × I	Depends on interest rate and duration of loan
Factors of annual sales revenue *S* or manufacturing (production) cost C_M		
Royalties	0–0.06 × S	Depends on need to purchase technology
Research and development (alternatively)	0–0.06 × S 0–0.05 × C_M	Depends on need for technology development within operating company
Selling expenses (alternatively)	0.02–0.2 × S 0.01–0.25 × C_M	Depends on nature of product (s), their uses, and number of customers
Corporate administration Cost of planning and managing business	0.02–0.06 × C_M	Depends on company policy, size, complexity, and nature of business

4.16 Effect of inflation on operating costs

It is frequently necessary to forecast future operating costs in dollars of the day, either as part of project evaluation or financial planning activities. It is also frequently desirable to update older operating cost data for the purpose of assisting current estimates. As discussed, operating costs comprise raw materials, energy, personnel, capital related and freight costs, and their relative contributions depend on the process technology and operating cost structure. Since the inflation rates for these contributing costs can move so independently, the use of a single inflation index, or even a uniformly weighted composite index (as used with plant capital costs), is inappropriate. The consumer price index (CPI), being focussed on costs of living, is also clearly inappropriate as an operating cost inflation index.

Movements over time in wages and salaries can be linked to historic government statistics for earnings. Maintenance costs may be more reasonably related to trends in plant cost indices since both costs are roughly equally divided between material and labour costs. Freight costs are more complicated, depending on trends in wages, capital, energy, technology, and work practices. Cognizance also needs to be made of learning and technological factors (discussed in Chapter 7), which exert a downward influence on many costs.

Dealing with inflation in project evaluation is discussed further in Chapter 5. Clearly, this becomes a higher priority in environments of higher inflation rates. Most operating companies and contractors will have forecasts of various cost movements including the CPI over a future period of say 3–5 years. The forecasts are likely to be limited in accuracy, but are still vital for planning. A rational basis devised for forecasting is also useful in the context of monitoring and understanding cost trends as they occur.

4.17 Scale influences on production costs

A convenient basis for examining the effects of scale on production costs is the simplified model outlined in Eq. (4.1). Of the performance parameters, raw materials consumptions, and utilities consumptions per tonne of product (R and E) are generally independent of capacity whilst numbers of personnel employed and fixed capital-dependent costs per tonne of product (M/QU and kI/QU) generally decrease with increased plant capacity.

There are some cases where energy consumption per tonne of product can depend on scale; high- or low-temperature processes can involve large heat losses to ambient air, and the surface area of equipment per unit of production capacity becomes influential. This effect is observable, for example, in small cryogenic air separation plants as well as high-temperature metallurgical reactors.

As discussed under Section 4.4.2, the number of personnel employed depends on the scale of the plant and is more heavily dependent for parallel streamed plants than for single streamed plants.

The dependence of fixed capital-dependent costs on capacity is not reported in the literature, though it is widely assumed that fixed capital-dependent costs maintain a fixed proportionality relationship with fixed capital. Thus it is a reasonable approximation to assume that the capacity exponent for fixed capital-dependent costs has the same value as for fixed capital investment (see Chapter 3, Section 3.4.1)

Unit costs of raw materials and energy may be different for different plant size options or they may be scale-invariant. For example, if the raw material is an intermediate from an upstream processing plant, then its production cost will benefit from scale economies and this may be passed on through its selling price. Plants consuming large quantities of raw material may be able to negotiate a discount. In other cases there may be a limited quantity of cheap feedstock available, with the next parcel more expensively priced so that unit costs effectively increase with scale. Similar considerations apply to energy or utility costs.

Suppose that for a given set of plant capacity options for a particular technology, the unit costs r, g, m defined in Eq. (4.1) are scale-invariant and that the factor k is also scale-invariant. At the same time, assume that R and E are scale-invariant whilst M and I are scale dependent. The dependence of production cost on capacity is then governed by the dependence of M and I on Q, the cost structure, or the dominance of Mm/QU and kI/QU in production cost.

The result may be relatively steep or flat scale curves [22]. Fig. 4.3 shows scale curves for ethylene, caustic soda/chlorine, and aluminium smelting plants from

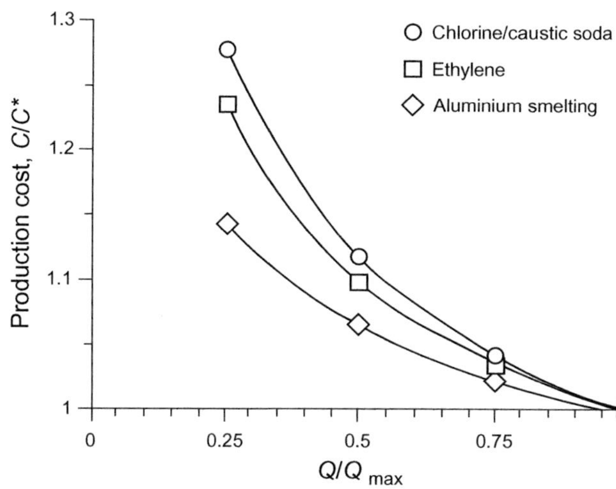

FIG. 4.3

Scale curves for ethylene, chlor-alkali, and aluminium smelting plants [22]. $C =$ production cost ($/t), $C^* =$ production cost ($/t) at Q_{max}, $Q =$ production capacity (t/year), and $Q_{max} =$ maximum available capacity (t/year) for the technology concerned.

Source: D.J. Brennan, Evaluating scale economies in process plants: a review, Trans. Inst. Chem. Eng. 70(A5) (1992), 516–526.

earlier studies by the author [22]. Whilst the aluminium smelter has a higher predominance of parallel streaming than the chlor-alkali and ethylene plants (reflected in the relative capital cost versus capacity exponents), the high contribution of variable cost arising from electricity consumption lowers the scale curve for aluminium smelting relative to the other technologies.

4.18 Calculating and presenting overall plant operating costs

Operating cost sheets are useful as checklists for ensuring all relevant costs are included and for presenting an approximate or more detailed estimate. Computer spreadsheets are ideal for estimating operating costs, for comparing alternative processes, for comparing alternative design options, for performing sensitivity studies, and for exploring effects of scale and capacity utilisation. Spreadsheet packages can also be used for linking individual worksheets to estimate overall operating costs where several plants make up a processing chain. Spreadsheets also enable production cost estimates to be linked directly with cash flow projections.

Table 4.1 showed an example of an operating cost sheet prior to completion. Table 4.7 shows an example of a simplified cost sheet used for a preliminary estimate. Table 4.7 is intended for illustrative purposes rather than an accurate definition of the actual cost of acetic acid manufacture. Cost sheets as depicted in Tables 4.1 and 4.7 can be more rigorously defined for detailed estimates, and provide a basis for

Table 4.7 Example of simplified operating cost summary sheet.

Product	**Acetic acid**			
Process route	Methanol carbonylation			
Plant capacity	40,000 t/year			
Capacity utilisation				
Product selling price				
Fixed capital	$100 million			
Production cost	Unit usage (/t product)	Unit cost ($/unit)	**Annual cost ($million)**	**Cost per tonne product ($/t)**
Raw materials				
Methanol	0.54	300		162
Carbon monoxide	0.52	300		156
Total raw materials cost				318
Utilities				
Electricity	0.12 MWh	$120/MWh		14.4

Table 4.7 Example of simplified operating cost summary sheet—cont'd

Fuel Steam Cooling water Other	2.5t $80 m^3$	\$16/t \$0.15/m^3		40 12
Total utilities cost				66.4
Process labour	**Number**	**Salary (\$/year)**		
Operators per shift	**4**	**\$70,000**		
Number of shift teams	**4.5**			
Total shift operators	**18**			
Day operators	**Nil**			
Total process labour Wages (*L*)			1.26	31.5
	% *L*			
Payroll overheads	40		0.50	12.6
Plant overheads	120		1.50	37.8
	% Fixed capital			
Maintenance	3.0		3.0	75.6
Insurance	1.2		1.2	3.0
Property taxes	1.0		1.0	2.5
Book depreciation	10.0		10.0	250
Total fixed cost				
Total production cost (TPC)		Fixed Variable Total		418.0 384.4 802.4
	% TPC			
Corporate administration	4.0			32.1
Research & development	2.0			16.0
Selling expenses	1.5			12.0
Total operating cost				862.5

exploring the effects of varying performance and unit cost inputs on the estimated production cost or overall operating cost. They are also ideal for reviewing the dominant cost contributions to overall operating costs.

4.19 Data sources for operating costs

Performance data for consumption of raw materials and utilities, as well as personnel requirements, may be drawn from a spectrum of literature sources dealing with specific technologies. Data sources may be classified under

- Public domain
  - Encyclopaedias of chemical technology
  - Journals
  - Books
- Classified
  - For example, in reports published by commercial consultants
- Computer packages
- Published information by operating companies

Some examples have been listed under Section 4.3 in the context of raw materials consumption. Public domain and classified sources for capital cost data, outlined in Chapter 3, are required input for capital-dependent costs. Sources of unit costs of raw materials have been discussed in Section 2.5, and of utilities in Section 4.3.2. Estimates of personnel wages and payroll overheads may be assisted by reference to labour statistical data as discussed in Section 4.5.

References

[1] Kirk-Othmer, Encyclopaedia of Chemical Technology, fifth ed., John Wiley, New York, NY, 2004.
[2] Ullmann's, Encyclopaedia of Industrial Chemistry, Electronic Release, Wiley, Weinheim, Germany, 2012.
[3] J.J. McKetta, Chemical Processing Handbook, Marcel Dekker, New York, NY, 1993.
[4] D.F. Rudd, S. Fathi-Afshar, A.A. Trevino, M.A. Stadtherr, Petrochemical Technology Assessment, John Wiley & Sons, New York, NY, 1981.
[5] J.H. Gary, G.E. Handwerk, M.J. Kaiser, Petroleum Refining. Technology and Economics, fifth ed., CRC Press, Boca Raton, FL, 2007.
[6] Australian Water Association, Desalination Fact Sheet, https://www.awa.asn.au/Documents/Desalination_Fact_Sheet.pdf. (Accessed 1 June 2019).
[7] J.R. May, D.J. Brennan, Sustainability assessment of Australian electricity generation, Trans. IChemE Part B 84B (2) (2006) 131–142.
[8] European Commission, Energy Prices and Costs in Europe 2018, January 2019 and related website. https://ec.europa.eu/energy/en/data-analysis/energy-prices-and-costs.
[9] D.J. Brennan, Aluminium smelting: a case study in modelling technological progress, Process Econ. Int. 7 (1) (1988) 30–41.

[10] D.J. Brennan, Capacity and Performance Profiles of Process Plants, (Ph.D. thesis), University of Melbourne, Melbourne, Australia, 1986.
[11] Comalco Ltd, Primary Aluminium Smelter. Summary of Proposed Development, Company Publication on Boyne Island Smelter, Brisbane, Australia, 1978.
[12] Alcoa Portland, Aluminium Smelter, 'Environmental Effects Statements and Draft Environmental Impact Statement', January, (1980).
[13] M.R. Wynn, G.H. Rutherford, Ethylene's unlimited horizons, in: European Chemical News Large Plants Supplement, 16 October, 1974, pp. 36–37.
[14] Department of Northern Development (Australia) and Department of Industrial Development (WA), The Pilbara Study (Australian Government Publishing Service), (1974).
[15] G. Guglielmi, Advanced technology in chlorine and caustic soda production plants, in: IChemE Symposium Series No. 50, IChemE, Rugby, UK, 1976, pp. D1–D10.
[16] G.R. Clews, Replacement of mercury chlor-alkali plants with new membrane plants in Australia, (Chapter 10). in: J. Moorhouse (Ed.), Modern Chlor-Alkali Technology, vol. 8, Blackwell Publishing, Abingdon, UK, 2001.
[17] US Bureau of Labor Statistics, International Labour Comparisons, August 2013.
[18] The Conference Board, International Labor Comparisons, https://www.conference-board.org/ilcprogram/.
[19] Victorian Government Statement, The Portland aluminium smelter, Economic Strategy for Victoria, Statement No. 3, August, 1984.
[20] E.E. Ludwig, Applied Project Engineering and Management, second ed., Gulf Publishing, Houston, TX, 1988.
[21] S. Mannan, F. Lees, Loss Prevention in the Process Industries: Hazard Identification, Assessment and Control, fourth ed., Elsevier, Woburn, MA, 2012.
[22] D.J. Brennan, Evaluating scale economies in process plants, Trans. Inst. Chem. Eng. 70 (Part A) (1992) 516–526.

CHAPTER

5 Evaluation of project profitability

We have heard it said that five percent is the natural interest of money.
Thomas Babington, 1st Baron Macaulay, English politician and historian (1800–59)

Chapter outline

5.1 Introduction

A fundamental requirement of any company or business is that it operates profitably. This means that after all operating costs are deducted from incoming revenue derived from the business activity, the company still generates a positive income which constitutes an adequate return on the assets employed. Otherwise, from a financial perspective, the money invested in those assets would be better consigned to a bank or other financial institution to earn interest.

All companies declare their profits to their shareholders in their annual report in the form of a profit and loss account. An illustrative example of a hypothetical process industry company is shown in Table 5.1. In this case, the operating revenue for the year was \$19.3 billion, which included \$18.6 billion in sales revenue. An operating profit of \$4.6 billion was declared before depreciation, amortisation, and interest. An amount of \$1.5 billion was charged for depreciation and amortisation.

Book depreciation charges based on established accounting procedures, depend on assumptions regarding the lives and values of assets. Many of these assumptions

Process Industry Economics. https://doi.org/10.1016/B978-0-12-819466-9.00005-5

Table 5.1 Example of an annual profit and loss account for a process industry company.

	$billion
Operating revenue and costs	
Sales revenue	18.6
Other revenue	0.7
Total revenue	19.3
Operating costs	14.7
Operating profit, including abnormal items, before depreciation, amortisation and interest expense	4.6
Deduct	
Depreciation and amortisation	1.5
Interest expense	0.4
Operating profit before income tax	2.7
Deduct	
Income tax expense including abnormal items attributable to operating profit	0.8
Operating profit after income tax	1.9
Deduct	
Outside equity interests in operating profit after income tax	0.2
Operating profit after income tax, attributable to operating company	1.7
Retained profits at the beginning of the financial year	5.5
Total funds available for appropriation	7.2
Dividends provided for or paid	0.7
Retained profits at the end of the financial year	6.5

may be debated; for example, the original cost of an asset adjusted for inflation may be quite different from its replacement value; an asset may still be in long use after its originally perceived life expectancy. Thus the measurement of profit is imprecise. There is also a fundamental distinction between how economists and accountants view profit [1].

At the evaluation phase of a project, a justifiable expectation of profitability is vital to a project's prospects of authorization. A simple and widely used indicator of profitability is the return on investment (ROI) defined as

$$\begin{aligned} \text{ROI} &= \frac{\text{Annual operating profit}}{\text{Total investment}} \\ &= \frac{\text{Annual sales revenue} - \text{Annual operating costs}}{\text{Fixed capital} + \text{working capital}} \end{aligned} \tag{5.1}$$

Sales revenue and operating costs are estimated at the estimated annual production level of a plant; sales revenue, operating costs, and investments are estimated in consistent currencies, usually of the year of evaluation. ROI is usually expressed as a percentage.

This is a before-tax measure where operating costs include book depreciation. Since return on investment can also be evaluated on the basis of after-tax cash flows, the return on investment measure should always be qualified, in particular whether before- or after-tax measure and whether including or excluding depreciation. Required returns on investment vary with the nature and risk of a proposed project. As a guide, however, for a process industry project of moderate risk with an expected life of 10–15 years, a threshold level of acceptance would be approximately 20% per annum ROI, before tax, and where operating costs include depreciation.

Table 5.2 shows a cost summary of alternative routes for producing hydrogen using natural gas or coal as the feedstock. The data in the table is intended as an approximate guide only for illustrating the use of simplified ROI assessments for profitability comparisons. The coal as a raw material has cost advantages compared with gas but incurs a higher capital cost and higher fixed operating costs excluding external environmental costs.

Whilst this simple ROI measure is widely used as a quick and useful indicator of profitability, the large capital investments with their associated risks and the complex nature of projects demand a more rigorous and sophisticated approach. The approach almost universally adopted by industrial companies and financing bodies is based on cash flow analysis. It provides a means of accounting for the duration and pattern of investment and the duration and pattern of profitable operation. The effects of capacity utilisation (often restricted in the initial operating years by limitations in market demand and/or plant reliability), of tax payments, selling price erosion, and inflation can all be accounted for. Cash flow forecasts assist in the planning of capital expenditure and hence finance requirements of projects can be amalgamated into a company's total financial plan.

Table 5.2 Profitability comparisons of hydrogen production from natural gas and from coal.

Feedstock	Natural gas	Coal
Plant capacity (tonnes/year)	**118,000**	**118,000**
Total fixed capital ($million)	**284**	**657**
Annual operating costs ($million)		
Raw materials	88.0	33.3
Utilities	14.1	42.1
Fixed costs	18.7	45.0
Nonmanufacturing costs	9.0	12.3
Total	**129.8**	**132.7**
Annual sales revenue ($million)		
Hydrogen sales	138.0	138.0
By-product steam	42.8	50.6
Total	**180.8**	**188.6**
Annual profit $ million	**51.0**	**55.9**
Return on investment % per annum	**18.0**	**8.5**

5.2 Cash flow estimation

Cash flow is the net flow of money into or out of a project (or a company) over a given time period. The commonly used time period in the economic evaluation of a project is 1 year. Cash outflows (e.g. capital expenditure on plant, expenditure on increased raw material stocks, tax payments) are considered negative, whilst cash inflows (e.g. income after tax, recovery of working capital at the end of project life) are considered positive.

Depreciation is not a cash flow. It is an internal allowance by a company to assist in providing for future investment in response to the erosion in the value of assets. A tax depreciation allowance, however, is a permissible deduction from income for tax assessment, and as such influences the magnitude of tax payments, and hence of after-tax cash flows.

Cash flows can be accumulated over the life of the project and represented on a 'cumulative cash flow diagram' or 'cash profile', see, for example, Fig. 5.1. The cumulative cash flow (also referred to as 'net cash position') can then be read directly from the diagram. Cash flows are calculated incorporating tax payments and are thus designated as after-tax cash flows. The tax situation can be complex, varying with changes in government policy and from one country to another, but the following aspects are important:

- Corporate tax is normally paid at a given rate of taxable income.
- Tax depreciation allowance, expressed as a percentage of fixed capital expended, is an allowable deduction from the operating cash flow (before tax) to arrive at taxable income. If the project life is less than the depreciable life (as determined by the tax depreciation rate) the residual balance can be claimed in the final year

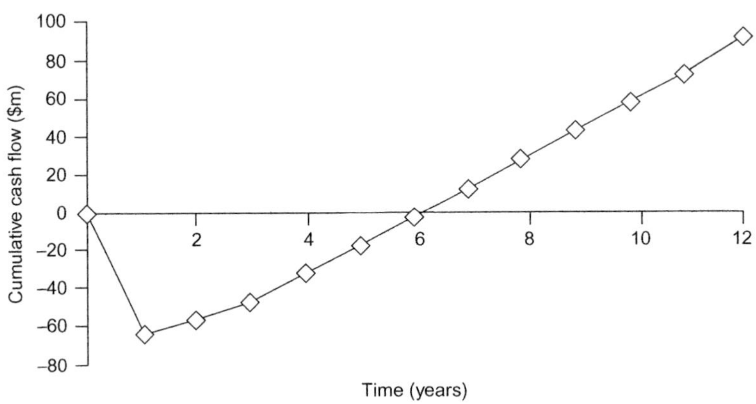

FIG. 5.1

Cumulative cash flow diagram.

of operation. The tax depreciation allowance is less for buildings than for plant, reflecting the potentially longer life of buildings.

- Periodically, governments may offer incentives to stimulate capital investment. One approach adopted is an investment allowance scheme. The investment allowance can generally be claimed once only in the first year of operation as a percentage of investment into the plant. The investment allowance, along with the tax depreciation allowance, is deducted from the operating cash flow before tax to arrive at taxable income.
- Tax payments may occur some time after income is incurred. In order to assist the management of their national economy, most governments will attempt to avoid any time delay, but some governments may allow delays for certain projects in order to encourage investment.
- In some project evaluations, the taxable income for a year may be negative. This is often the case in the first year of operating life for a project. If the project can be considered as one of a portfolio of investments which are collectively profitable, a tax credit for that year can be assumed for the project concerned. Provision may also be made under the tax rules to carry the loss into a future year for assessment.
- Government policy on taxation can vary considerably from year to year within a given country and can also vary considerably between countries. This applies to the whole spectrum of corporate tax including the tax rate, depreciation allowances, the timing of tax payments, special incentives, and so on. The corporate tax rate for Australia was 30% in 2019 but was as high as 49% in 1986.
- Corporate tax rates for different countries in 2019 ranged from as low as 17% to as high as 35%. Some countries may allow very rapid tax depreciation schedules, some may provide an investment grant (as distinct from an investment allowance) as a percentage of fixed capital, and some may provide for tax 'holidays' waiving the requirements to pay tax for some years after project commencement. Details of tax rates in different countries are published in financial reports by national governments.

To illustrate cash flow estimation, a simple example 5.1 is provided.

Example 5.1

A process plant of capacity 100,000 tonne/year of product 'Magidrop' is to be constructed next year at a fixed capital cost of $50 million, to be ready for operation at the beginning of the following year. Sales volume is forecasted to be 70,000 tonne/year in the first year of operation, increasing to 85,000 tonne/year in the second year, and 100,000 tonne/year in third and subsequent years. The following economic data package and tax assumptions apply; these have been selected solely to illustrate the principles involved in cash flow estimation.

Economic data package	
Plant production capacity	100,000 tonne/year
Fixed capital cost	$50 million spent over 1 year

Working capital	Negligible
Product selling price	$1000/tonne
Variable operating costs	$700/tonne
Fixed operating costs (excluding depreciation)	$8 million/year
Sales volume	70,000 tonne/year in the first year of operation
	85,000 tonne/year in the second year of operation
	100,000 tonne/year in the third and each subsequent year of operation
Plant operating life	10 years

Tax assumptions	
Corporate tax rate	40%
Tax depreciation rate	10%/annum until fully depreciated
Tax payments	made in the year in which income occurs

A cash flow table over the life of the project is drawn up—see Table 5.3. Since fixed costs exclude depreciation, the before-tax cash flows during plant operating years are equal to the sales revenue less fixed and variable costs. The before-tax cash flows correspond to the term EBIT (earnings before interest and tax) which is widely used in accounting and business circles. The after-tax cash flows are equal to the before-tax cash flows which is less than the tax paid. Note that in the cash flow table, only rows a, b, c, d, e, and f are cash flows. If the tax payment were expressed as project cash flow its sign would be reversed. The final two rows of the table summarise the after-tax cash flows on an individual and a cumulative basis.

Example 5.2

In Example 5.1, it was assumed that working capital was negligible. The case is now explored where working capital requirements are equivalent to 20% of annual sales revenue. Thus at the end of year 1 (or beginning of year 2) working capital requirements are $14.0 million. At the end of year 2 requirements are $17.0 million, and at the end of year 3 (and until the end of year 11) are $20.0 million.

Expressed as cash flows these become (in $million)

Year	**1**	**2**	**3**	**4–11**	**12**
Working capital	−14.0	−3.0	−3.0	0	+20.0

and the after-tax cash flows become (in $million)

Year	**1**	**2**	**3**	**4–11**	**12**
After-tax cash flows	−64	+6.8	+9.5	+14.2	+20.0

See Table 5.4.

Table 5.3 Cash flow table for Example 5.1. Cash flows are in $million.

Year		1	2	3	4	5	6	7	8	9	10	11
Fixed capital		−50										
Working capital		0										
Sales volume	kt/year		70	85	100	100	100	100	100	100	100	100
Selling price	$1000/t											
Sales revenue			70	85	100	100	100	100	100	100	100	100
Variable costs	$700/t											
			−49	−59.5	−70	−70	−70	−70	−70	−70	−70	−70
Fixed costs			−8	−8	−8	−8	−8	−8	−8	−8	−8	−8
Cash flow before tax			**13**	**17.5**	**22**	**22**	**22**	**22**	**22**	**22**	**22**	**22**
Tax depreciation rate (%/year)	10											
Tax depreciation allowance			5	5	5	5	5	5	5	5	5	5
Taxable income			8	12.5	17	17	17	17	17	17	17	17
Tax rate (%)	40											
Tax payment			−3.2	−5	−6.8	−6.8	−6.8	−6.8	−6.8	−6.8	−6.8	−6.8
Cash flow after tax		**−50**	**9.8**	**12.5**	**15.2**	**15.2**	**15.2**	**15.2**	**15.2**	**15.2**	**15.2**	**15.2**
Cumulative cash flow after tax		**−50**	**−40.2**	**−27.7**	**−12.5**	**2.7**	**17.9**	**33.1**	**48.3**	**63.5**	**78.7**	**93.9**

Table 5.4 Cash flow table for Example 5.2. Cash flows are in $million.

Year		1	2	3	4	5	6	7	8	9	10	11	12
Fixed capital		−50											
Working capital		−14	−3	−3									20
Sales volume	kt/year		70	85	100	100	100	100	100	100	100	100	
Selling price	$1000/t												
Sales revenue			70	85	100	100	100	100	100	100	100	100	
Variable costs	$700/t		−49	−59.5	−70	−70	−70	−70	−70	−70	−70	−70	
Fixed costs			−8	−8	−8	−8	−8	−8	−8	−8	−8	−8	
Cash flow before tax			**13**	**17.5**	**22**	**22**	**22**	**22**	**22**	**22**	**22**	**22**	
Tax depreciation rate (%/year)	10												
Tax depreciation allowance			5	5	5	5	5	5	5	5	5	5	
Taxable income			8	12.5	17	17	17	17	17	17	17	17	
Tax rate (%)	40												
Tax payment			−3.2	−5	−6.8	−6.8	−6.8	−6.8	−6.8	−6.8	−6.8	−6.8	
Cash flow after tax		**−64**	**6.8**	**9.5**	**15.2**	**15.2**	**15.2**	**15.2**	**15.2**	**15.2**	**15.2**	**15.2**	
Cumulative cash flow after tax		**−64**	**−57.2**	**−47.7**	**−32.5**	**−17.3**	**−2.1**	**13.1**	**28.3**	**43.5**	**58.7**	**73.9**	**93.9**

Notes: (1) Although the working capital commitment increases from $14.0 million to $17.0 million and $20.0 million, the corresponding cash flows represent the incremental changes to working capital. (2) Working capital is not treated as an operating expense for taxation purposes. Hence the working capital cash flow in year 2 for example is added as a negative cash flow to the operating cash flow after tax. If fixed capital expenditure occurred in year 2 this would likewise be added as negative cash flow, although it would attract tax allowances. (3) The influence of working capital is to change the cumulative cash flow pattern, though since working capital is deemed recoverable, the cumulative cash flow at the end of project life is the same in Example 5.2 as in Example 5.1.

5.3 Measures of profitability

Based on the cumulative cash flow diagram, a number of derived measures of profitability can be used, measured in terms of time, cash, and percent return on investment.

Payback time is a term often loosely used to denote the time taken to recover investment costs. It is sometimes taken as the time from commencement of the project to recover the initial capital investment (land, fixed capital, and working capital). More frequently however it is taken as the time from the start of production to recover fixed capital expenditure only; working capital and land capital are considered as being recoverable at the conclusion of the project. Fig. 5.2 illustrates these alternative interpretations.

When measuring profitability, we need to account for the time value of money. A sum of money can change in value over time due to two separate causes:

(i) earning capability
(ii) inflation

The two causes are distinct in concept and measurement, though both must be taken into account in an investment proposal. Initially, for clarity, only the earning capability of money is considered, assuming a business environment of negligible inflation.

Money invested has earning capability expressed as a rate of interest per annum, i, ranging from a lower rate on low-risk investment to a higher rate for greater risk. Examples of low-risk investments include bank savings accounts or long-term government bonds. An example of a high-risk investment would be the purchase of shares in a company exploring for diamonds in a politically unstable country. Interest has often been subdivided into a risk-free component and a risk premium. Over time, the money value V of an original sum S grows by the compounding of interest earned at rate $i\%$ per annum over time t (years) such that

$$V = S(1+i)^t \tag{5.2}$$

where i is expressed as a decimal.

Net present value is a measure of the net cash benefit generated by a project. It is built on the concept of present value, which is an approach used to bring the value of future cash flows to a consistent basis (in this case the present), taking into account the interest-earning capability of money. The present value of a future cash flow C_t occurring t years from the present is the value of a cash flow which, if invested now at an interest rate i would in t years time amount to C_t as a result of compounding interest. Thus:

$$\text{Present value} = C_t/[1+i]^t \tag{5.3}$$

This approach allows, for example, an expenditure cash flow of \$10 million in 2 years time, and a cash flow benefit of \$20 million dollars in 10 years time to be brought to a common measurable basis (at the present). If the interest rate is taken as 12%/annum, the present value of the expenditure cash flow becomes $-10/(1.12)^2$ or $-\$8.0$ million. The present value of the income cash flow becomes $+20/(1.12)^{10}$ or +\$6.4 million.

Alternatively, the concept of the future value of cash flow could be used. However, as the present is rather more definite than the future, and since the economic life of a project is invariably uncertain, the present value is generally accepted as the more convenient and practical concept.

Net present value (NPV) for a project is the net value of the present value of all cash flows for the project, from the commencement of capital expenditure to completion of economic life. Thus:

$$\mathrm{NPV(Project)} = \sum_{t=0}^{t=k} C_t/[1+i]^t \tag{5.4}$$

where k = total number of years over which cash flows occur.

NPV can be regarded as total income less total expenditure over the entire life of the project, where all incomes and expenditures are expressed as present values. Cash flows and interest rates need to be expressed using a consistent basis; this basis is most commonly and appropriately after tax.

Since most process industry projects involve an initial investment phase followed by a longer income-generating phase, the higher the value of i, the lower is the NPV for the project. A positive NPV implies that a net cash benefit is obtained as a result of the project; hence the project is profitable. A negative NPV implies a loss for the project. A zero NPV implies we have neither gained nor lost from the project; we have 'broken even', and the financial result is the same as if we had invested our funds at i% per annum over the project life.

Since cash invested today yields higher future values through investment earnings, future cash flows are reduced when brought to a present value basis; hence they are said to be 'discounted'. The value of i for which a given set of cash flows yields an NPV of zero is termed the discounted cash flow return (DCFR), or alternatively the internal rate of return (IRR). By definition, it is the value of i for which

$$\sum_{t=0}^{t=k} C_t/[1+i]^t = 0$$

and must be determined iteratively. If the DCFR earned by the project exceeds the interest rate payable on money invested and generated in the project, then the project is profitable. If the DCFR is less than that interest rate, the project is unprofitable.

5.4 Discount rate selection

The selection of the appropriate discount rate for NPV calculations is not a clear cut decision. The rate during the expenditure phases of the project should reflect the cost of funds borrowed or used for the project. If the funds are borrowed from an external source (loan funds), interest payments are deductible from monies earned for tax purposes, and the after-tax cost of money is $i(1-t)$ where i is the interest rate charged and t is the tax rate. If equity funds (generated from within the company) are used, these are deemed to have a value estimated by company accountants which reflect the earning potential of the funds if invested elsewhere at a comparable risk. This earning potential is termed opportunity cost. The cost of capital for the project is then the weighted cost of funds from the separate sources. The after-tax cost of loan funds is likely to be less than that for equity funds, but the loan funds are a greater financial liability for the investor since these must be repaid as interest plus principal.

The exact financing structure for a project will not be resolved until the authorisation phase, and hence both the borrowing rates and the relative contributions of loan and equity funds are uncertain during the development phase of a project. Not only does the discount rate adopted affect the apparent profitability of the project, but it can shape certain decisions such as the sizing and timing of investment and the balance between capital and operating costs. It is important to determine how sensitive these decisions are to the discount rate.

During project evaluation, the engineer (or person responsible for the evaluation) will need to liaise with accounting or financial personnel for guidance in discount rate selection and other financial details, for example, in relation to taxation. Liaison will increase as the project becomes more definitive. Because of inevitable uncertainty in the exact value of the discount rate during project development, it is important to estimate NPVs at several discount rates in the vicinity of the expected value.

Example 5.3. Calculating the indices of profitability

The cash flows estimated in Example 5.2 can now be used to illustrate the calculation of profitability indices. The cumulative cash flow diagram is first drawn (see Fig. 5.2). The shape of the diagram is a good first guide to project profitability. In this case, the indication is one of marginal profitability. The cumulative cash flow is positive for the greater part of the project life, and at the end of the project is some 50% greater than the peak cumulative debt.

Payback time, measured as the time from commencement of operation to recover fixed capital expenditure only is approximately 3.2 years. From the commencement of expenditure to the recovery of fixed and working capital expenditure is 5.6 years. These indicators suggest a project of marginal profitability, but they are short-term indicators. The remaining cash flows, the project life, and the cost of capital must be considered.

For the purpose of this exercise, the cost of capital is taken as 10% per annum. The cash flows are then discounted at this rate over the life of the project. For illustrative purposes, the cash flows discounted at 10% per annum and other rates are listed in Table 5.5. Table 5.6 summarises the discounted cash flows and NPV at the discount rate of 10% per annum.

The NPV, the sum of the discounted cash flows is $21.9 million, indicating the project is profitable. $21.9 million is the value in today's money which the project is estimated to yield if the cost of capital is 10% per annum.

The effect of the discount rate on the project NPV is now explored. Note that the NPV for $i=0$ corresponds to the cumulative cash flow for the project. The NPVs for various discount rates are summarised in Table 5.7 and illustrated in Fig. 5.3.

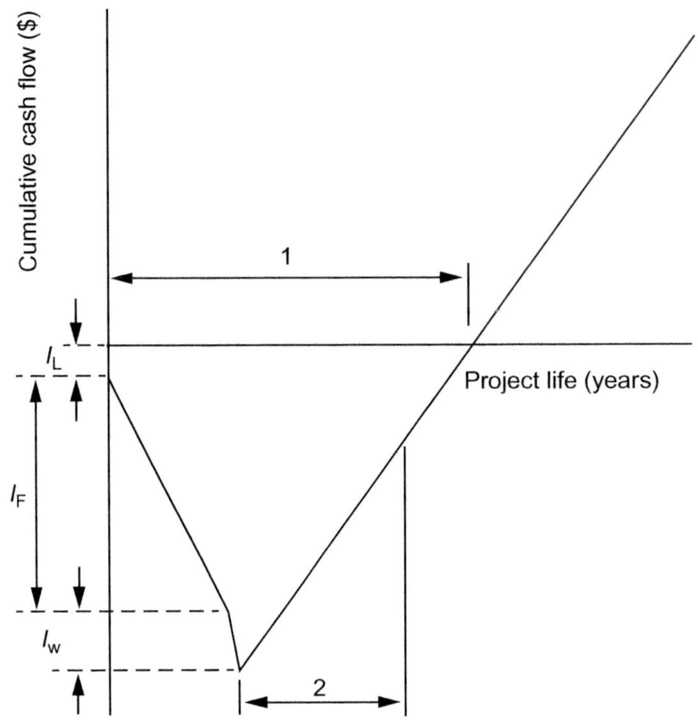

FIG. 5.2

Alternative definitions of payback time: I_L = land capital; I_F = fixed capital; and I_W = working capital.

The DCFR, the discount rate for which NPV = 0, is 16.6% per annum. Fig. 5.3 shows that the NPV versus discount rate curve is not linear, though approximate interpolations can be made with care.

An average return on investment over a project life of (93.9 + 64.0)/(64 * 11) or 22.4%/annum can be estimated from the cumulative cash flow diagram shown in Fig. 5.1 where the cumulative cash flows are those calculated in Example 5.2 and shown in Table 5.4. Note that this estimate is based on after-tax cash flows. For comparison, a simple before tax return on investment based on Eq. (5.1) assuming 10% depreciation of fixed capital per annum and full plant loading comes to 17.0/70% or 24.3% per annum.

5.5 Reviewing profitability indices

The indices of profitability are measured either in %/annum, cash, or time. Rather than adopt a single index of profitability, it is better to use all of the indices outlined, in conjunction with the cumulative cash flow diagram from which they are derived.

Table 5.5 Discounted cash flows from Example 5.2.

	Cash flow after tax, $million	Discounted cash flow after tax ($million)								
Year	**Discount rate/year**	**0**	**0.025**	**0.05**	**0.075**	**0.1**	**0.125**	**0.15**	**0.175**	**0.20**
1	−64.0	−64	−62.4	−61	−59.5	−58.2	−56.9	−55.7	−54.5	−53.3
2	6.8	6.8	6.5	6.2	5.9	5.6	5.4	5.1	4.9	4.7
3	9.5	9.5	8.8	8.2	7.6	7.1	6.7	6.2	5.9	5.5
4	15.2	15.2	13.8	12.5	11.4	10.4	9.5	8.7	8.0	7.3
5	15.2	15.2	13.4	11.9	10.6	9.4	8.4	7.6	6.8	6.1
6	15.2	15.2	13.1	11.3	9.8	8.6	7.5	6.6	5.8	5.1
7	15.2	15.2	12.8	10.8	9.2	7.8	6.7	5.7	4.9	4.2
8	15.2	15.2	12.5	10.3	8.5	7.1	5.9	5	4.2	3.5
9	15.2	15.2	12.2	9.8	7.9	6.4	5.3	4.3	3.6	2.9
10	15.2	15.2	11.9	9.3	7.4	5.9	4.7	3.8	3	2.5
11	15.2	15.2	11.6	8.9	6.9	5.3	4.2	3.3	2.6	2
12	20.0	20.0	14.9	11.1	8.4	6.4	4.9	3.7	2.9	2.2
Total	93.9	93.9	68.9	49.4	34.1	21.9	12.1	4.3	−2.0	−7.1

Table 5.6 Discounted cash flows and NPV at a discount rate of 10% per annum.

Year	Cash flow after tax ($million)	Discounted cash flow after tax ($million)
1	−64.0	−58.2
2	6.8	5.6
3	9.5	7.1
4	15.2	10.4
5	15.2	9.4
6	15.2	8.6
7	15.2	7.8
8	15.2	7.1
9	15.2	6.4
10	15.2	5.9
11	15.2	5.3
12	20.0	6.4
		Total 21.9 million

Table 5.7 Effect of discount rate on NPV.

Discount rate, % per annum	Net present value (NPV) $million
0	93.9
2.5	68.9
5	49.4
7.5	34.1
10	21.9
12.5	12.1
15	4.3
17.5	−2.0
20	−7.1

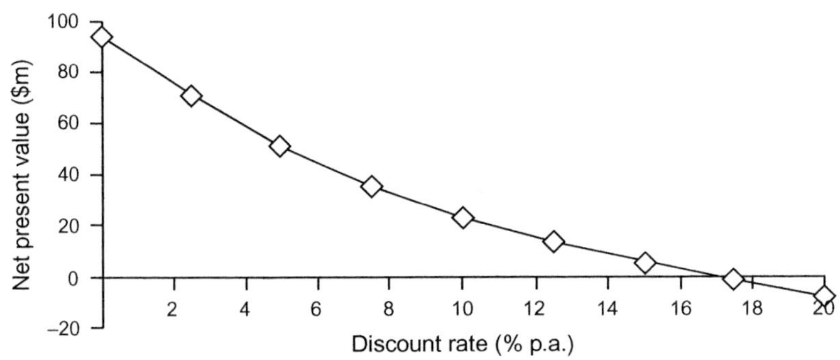

FIG. 5.3

Relationship between NPV and discount rate.

This approach encourages a complete and balanced view of the project profitability over its life incorporating investment, early operation and later operation phases. NPV and DCFR should always be estimated to ensure that the time value of money is accounted for. NPV measures the cash value of the project but requires knowledge of the cost of capital and needs to be considered in the light of the magnitude of the investment. For example, an investment of $50 million yielding an NPV of $20 million may well be preferred to an alternative investment of $100 million yielding an NPV of $24 million because of the smaller magnitude of capital risked. The DCFR, by comparison, does not require the cost of capital as an input, although eventually the DCFR value only has meaning if the cost of capital is available as a yardstick for comparison. The DCFR may also sometimes appear to conflict with the NPV as a profitability indicator as discussed below.

An advantage of NPV over DCFR as a profitability index is that the NPVs are additive whilst DCFRs are not. Where a large project can be broken down into individual components for analysis, it is convenient to analyse each component separately for profitability, but also desirable to amalgamate the cash flows and NPVs of the various components to evaluate the entire project. An example of a very large project would be olefins and derivatives project broken down into the olefines plant, the derivatives plants (such as polyethylene, polypropylene, styrene) and the related utility plants. Smaller projects can similarly be broken down into components for separate evaluation.

So far, examples of profitability indices have been used in the context of approving or rejecting a single project. The availability of capital for new plants or modifications to existing plants is often limited, and profitability indices are then used alongside other (nonfinancial) criteria to rank competing projects. The DCFR is widely used for this purpose, but in this context, as when examining a single project, the DCFR should be considered in conjunction with other indicators, especially the NPV. Fig. 5.4 shows a situation for two competing projects where the project with the higher DCFR has the lower NPV at the cost of capital. At the cost of capital indicated in Fig. 5.4, Project A will be the larger net generator of funds and therefore preferred.

Frequently, because of capital shortages, target rates of DCFR for projects are set well above the rates of return actually being earned by a company. These higher rates are then used as discount rates for forecast cash flows in projects. This policy leads to those projects with longer-term benefits being potentially penalised, since the cash flows generated later in the project are heavily discounted [2].

5.6 Cost minimisation

Sometimes projects must be implemented where it is difficult to quantify the benefits. Examples include modifications to plant for safety or environmental reasons or to accommodate a change in the feedstock. In such cases, it is not possible to include sales revenues and the estimated cash flows are negative. However, there may still be

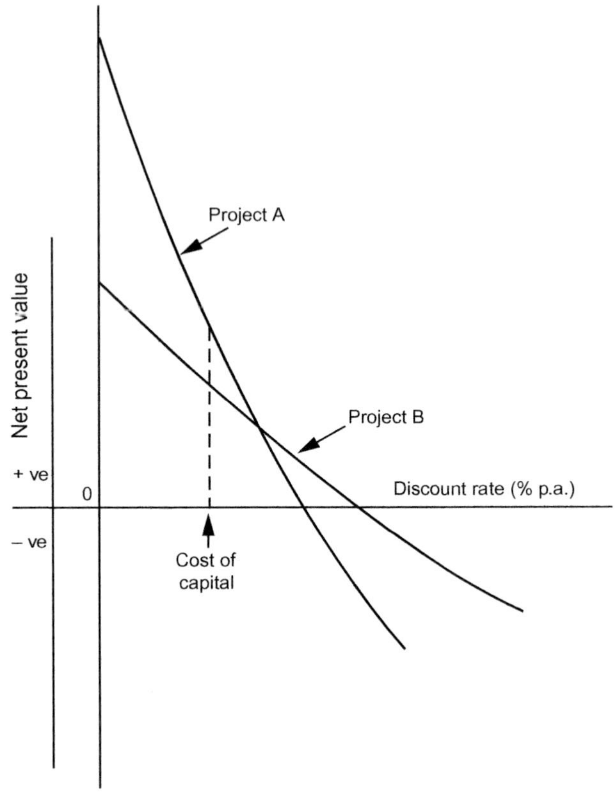

FIG. 5.4

NPV versus discount rate profiles for two competing projects.

alternative schemes for each project, for example alternative process routes for a waste treatment project. In such cases, the objective should be to minimise the net present cost for the project (that is, to minimise the negative NPV).

5.7 Dealing with inflation in cash flow evaluation

Inflation is the term used to describe the erosion in the purchasing power of a monetary currency with time. It can alternatively be thought of as the apparent escalation in cost over time of a given item or commodity of consistent quality.

In considering inflation, it is necessary to distinguish between nominal and real sums of money. If, for example, we invest \$100 at 10% per annum in a bank, we can withdraw the investment 1 year later as \$110 (assuming no bank fees!). The withdrawal represents a *nominal* sum of \$110. If, however, the rate of inflation over the same period is 10% per annum, the *real* value of our withdrawal with reference to our original investment will be $100. Thus we have gained no financial advantage from the investment. Alternatively, a government may promise to increase expenditure on health in the coming year's budget by 8% over the previous year. If, however, the rate of inflation is 10% per annum, the *nominal* increase of 8% implies a *real* reduction of 2% in health expenditure.

Different costs can be observed to escalate at different rates, but inflation is generally measured by an index of the cost of living, frequently referred to as the consumer price index (CPI). The consumer price index measures the change in the price of a 'basket' of goods and services in the areas of food, clothing, housing, household equipment, transportation, tobacco and alcohol, health care, recreation, and education, which are collectively deemed to comprise the 'cost of living'. CPI data for different countries are provided in Part 2 of the United Nations statistical yearbooks published annually [3].

The values of CPI for various countries are listed in Appendix 2. Whilst in years of economic depression there have been periods of deflation, there has generally been moderate and occasionally high inflation in most Western countries. The highest inflation rates were experienced in the 1970s; this was a period of rapid growth in the world price of crude oil. For example, the price of Arabian light crude oil increased from around US$2/bbl in the 1960s and early 1970s to $11.60 in 1974, and further to over $30/bbl in the early 1980s. This caused a general cost increase in other energy-based commodities and for hydrocarbons, which in turn translated into cost increases for hydrocarbon derived materials and ultimately a whole range of products. There were also increases in salaries and wages as communities struggled to afford the goods which contributed to their lifestyles. In the years from the early 1980s to the mid-1990s, the price of crude oil decreased, both in nominal and real terms to the point wherein 1995 in real terms, it was approximately twice its value in 1970.

Analysis of past trends in costs or prices shows clearly that the rates of change differ considerably from one item to another over a given time period. Changes in costs and prices relative to the change in CPI are designated as real changes. The real selling price of most petrochemicals in the US declined in the 1960s, but increased between 1974 and 1980 due to the rapidly escalating costs of feedstocks and energy sources derived from crude oil. Plant cost indices have, at certain times, moved ahead of CPI values reflecting escalation in real costs of materials and labour.

Thus there can be entirely different trends in a given country in cost escalation for capital, operating labour, raw materials, energy, freight, and the CPI, making cost accounting and forecasting difficult. Not only do overall rates of inflation measured by CPI differ for different countries, but rates of inflation in contributing costs can differ from country to country. For example, in the 1970s the real cost of salaries and wages increased in Australia and decreased in the United Kingdom.

Most companies use forecasts (typically 5 years) of inflation in the major elements of cost. Such forecasts, based largely on trends at the time of forecasting, are recognised to be uncertain but are necessary for planning.

Three distinct approaches to cash flow analysis have been used in inflationary environments.

Approach 1

Cash flows are estimated on the basis of costs and prices at the present, or the time of evaluation. This approach is justified by its users on the basis that inflation rates are highly unpredictable, and that their incorporation leads to complexity in analysis and computation. The disadvantages of such an approach are that relative rates of inflation for different costs (e.g. labour and energy) are not allowed for and that some elements (e.g. tax depreciation allowance) are held constant when in fact they decline in real value. Tax depreciation allowances for most countries, in fact, remain constant in nominal terms despite inflation.

Approach 2

Cash flows are estimated on the basis of future costs and prices. Estimates of inflation rates for various parameters contributing to cash flow such as energy, raw materials, labour, fixed capital, and product selling prices can be used to calculate inflated cash flows. Certain elements, for example, tax depreciation allowances, will remain uninflated over time.

This approach has the advantage of enabling different escalation rates to be taken into account, and of forecasting the cash flows in actual currency units of the year they occur. This is an essential part of financial planning to ensure funds are available to meet projected requirements. There is considerable difficulty however in forecasting inflation rates accurately, and there are potential difficulties in cash flow analysis. When one examines the table of projected cash flows, it is difficult to discern trends such as those resulting from changes in capacity utilisation or modes of operation, since these are obscured by inflation. Errors in data insertion may also be obscured.

Approach 3

Cash flows are estimated as in the second approach but are converted to a real basis by dividing by a deflator, such as a consumer price index. Thus inflation and uninflated tax depreciation allowances are accounted for, and the deflation permits easier comparison of annual cash flows. The deflation can be applied to entire cash flows or to components of annual cash flows such as sales revenues, variable operating costs, fixed operating costs, fixed capital, and so on. Whilst more irksome, the deflation of the components performs an important function in assisting the checking of the estimates.

It may be seen that the second approach estimates *nominal* cash flows and the third approach estimates *real* cash flows. The first approach by ignoring inflation might appear to estimate 'pseudo-' real cash flows but is fundamentally flawed as a rigorous approach to inflation. The first approach is suitable only for early stages of project development, or for cases where very low inflation rates apply.

5.7.1 Discount rate selection

In discounting nominal and real cash flows it is important to use the appropriate interest rate. In the same way that is it is important to distinguish between nominal and real cash flows, it is important to distinguish between nominal and real interest rates or rates of return.

As discussed by Wilkes [4], one can relate the real rate of interest 'e' to the nominal rate of interest 'i' by the expression

$$1+e=(1+i)/(1+\alpha) \tag{5.5}$$

where α = annual inflation rate, expressed as a decimal.

Thus if the nominal interest rate were 15%, the inflation rate 10%, the real interest rate would be 4.5%.

In an inflationary environment, the nominal cost of capital for a project can be estimated from a knowledge of equity and loan interest rates, and the funding structure for the project. This can then be applied to nominal cash flows to establish the net present value for the project using Eq. (5.3), where C_t would be the nominal cash flow in year t.

Similarly, the net present value could be established using real cash flows and the real discount rate:

$$\text{NPV}=C_t/(1+e)^t \tag{5.6}$$

where C_t = real cash flow in year t.

The net present values obtained from Eqs (5.4), (5.6) are seen to be equivalent, since by definition

$$C/C=1+\alpha$$

and $(1+i)/(1+e)=1+\alpha$ [from Eq. (5.5)].

Whilst it is correct to use nominal interest rate i with nominal cash flows C, or real interest rate e with real cash flows C for net present value calculation, it is clearly inappropriate to use nominal interest rate i with real cash flows C.

It is also important to recognise that a DCF return estimated from nominal cash flows is a nominal DCF return; this can be brought to a real basis using Eq. (5.5).

Eq. (5.5) enables real interest rates to be calculated from a knowledge of nominal interest rates and inflation rates and applies in any economic environment. The value of the real rate varies with the economic environment.

Government bonds provide a useful example of low-risk investment. An analysis of real interest rates on long-term bonds in the United States between 1790 and 1980 by Leuthold [5] indicated that real interest rates tended to be lower in periods of higher inflation. When annual inflation ranged between 0.5% and 1.4%, the real interest rate averaged 3.3% per annum; at inflation levels of 1.5%–2.4%, the real rate declined to an average of 2.1% whilst at inflation levels over 7.5%, the real rate declined to a negative 6.7%. Fig. 5.5 shows the movement in real interest rates of long-term government bonds in Australia. From 1949 to 1979, the Australian trend was similar to that reported in the US study. In the period 1980–87 in Australia

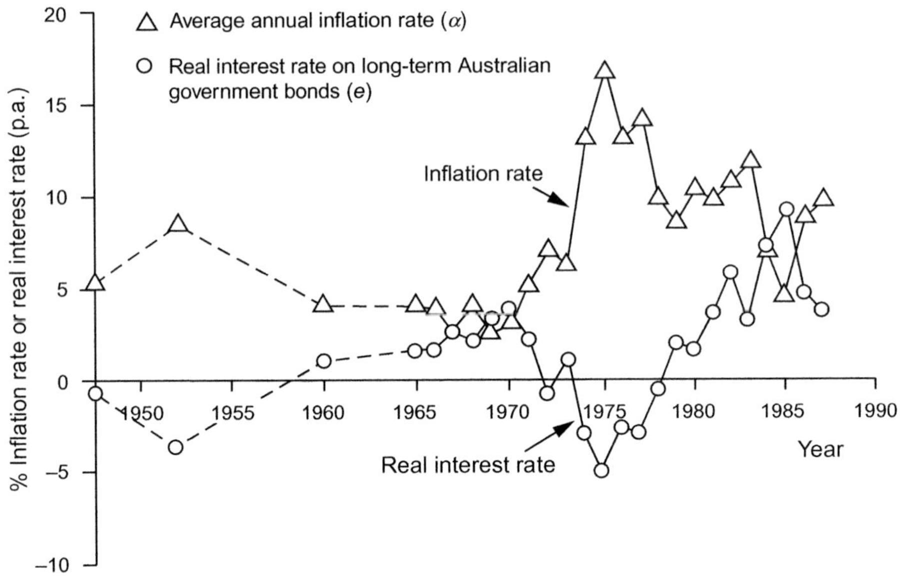

FIG. 5.5

Some historic data on real interest rates on Australian long-term government bonds.

however, real interest rates were relatively high during periods of moderately high inflation. This collective evidence confirms the view that the real rate of interest varies considerably with time and the prevailing economic conditions.

Example 5.4. Cash flow estimation and discounting in an inflationary environment

The effects of the various approaches to inflation in cash flow evaluation are now explored using the problem given in Example 5.1. A moderately high inflation rate is assumed with a nominal after-tax cost of capital of 16% per annum, and specific annual escalation rates as follows:

Fixed capital	12%
Selling price	9%
Variable costs	11%
Fixed costs	10%
Consumer price index	10%

Using Approach 1, the after-tax cash flows are as estimated in Example 5.1. In that example, inflation was ignored. In this example, inflation is taken as 10%/annum and the nominal after-tax cost of capital as 16%/annum. Thus the real cost of capital e is obtained from Eq. (5.5)

$$e = (1+i)/(1+\alpha) - 1 = 0.055 \quad \text{or} \quad 5.5\% \text{ per annum}$$

Table 5.8 shows the resulting discounted cash flows for each year of the project and the corresponding NPV and DCFR.

Table 5.8 Discounted cash flows and corresponding project NPV and DCFR.

Year	Cash flow after tax ($million)	Present value of after tax cash flow ($million)
1	−50.0	−47.4
2	9.8	8.8
3	12.5	10.65
4	15.2	12.27
5	15.2	11.63
6	15.2	11.02
7	15.2	10.45
8	15.2	9.90
9	15.2	9.39
10	15.2	8.90
11	15.2	8.43
Project net present value	**$54.1 million**	
DCFR	**23.9%/annum**	

Using **Approach 2**, cost and price components of cash flow are escalated as shown in Table 5.9. All sums and cash flows are at $million and are nominal. Cash flows are discounted at the nominal cost of capital.

Note that the annual tax depreciation allowance based on the fixed capital actually expended remains constant in nominal terms but declines in real terms. Annual fixed costs increase due to inflation, whilst annual sales and variable costs increase due to both inflation and increasing production and sales. Scanning through the cash flow table, however, the trends in sales, costs, and cash flow after tax is not readily discerned.

Using Approach 3, cash flows after tax are estimated as in Approach 2 but are brought to a real basis, using the consumer price index as the deflator in part (b) of Table 5.9. The cash flows are then discounted using the real cost of capital of 5.5%/annum. The present values of cash flows are seen to be identical to those calculated using Approach 2.

Approach 3 could be extended further to bring individual components of cash flow to a real basis if this were considered warranted.

Note that the real cash flows after tax and hence project profitability are far less encouraging than those estimated in Approach 1. This is due to the proper treatment of tax depreciation allowances, but also due to the assumptions on the relative movements of the selling price, cost, and CPI. There are strong grounds for anticipating declines in real selling prices over project life due to the effects of competition and technological obsolescence as discussed in Chapters 2 and 7. It is not uncommon, however, for this effect to be ignored in project evaluation. Note that for the relative escalation rates assumed, we have a situation where there is a steady decline in the annual before-tax cash flows even in nominal terms, after year 4.

Table 5.9 Cash flow table for Example 5.4 with inflated price and cost inputs.

Nominal cash flows ($million)	**Current value**	**Inflation rate (%/year)**											
Year			**1**	**2**	**3**	**4**	**5**	**6**	**7**	**8**	**9**	**10**	**11**
Fixed capital		12	−56										
Working capital			0										
Sales volume (kt/year)				70	85	100	100	100	100	100	100	100	100
Selling price ($/t)	1000	9											
Sales revenue				83.2	110.1	141.2	153.9	167.7	182.8	199.3	217.2	236.7	258
Variable costs ($/t)	700	11											
Variable costs				−60.4	−81.4	−106.3	−118	−130.9	−145.3	−161.3	−179.1	−198.8	−220.6
Fixed costs	8	10		−9.7	−10.6	−11.7	−12.9	−14.2	−15.6	−17.1	−18.9	−20.7	−22.8
Cash flow before tax				13.1	18.1	23.2	23	22.6	21.9	20.8	19.3	17.2	14.6
Tax depreciation rate (%/year)	10												
Tax depreciation allowance				5.6	5.6	5.6	5.6	5.6	5.6	5.6	5.6	5.6	5.6
Taxable income				7.5	12.5	17.6	17.4	17	16.3	15.2	13.7	11.6	9
Tax rate (%)	40												
Tax payment				−3	−5	−7	−7	−6.8	−6.5	−6.1	−5.5	−4.7	−3.6
Cash flow after tax			−56	10.1	13.1	16.1	16.1	15.8	15.4	14.7	13.8	12.6	11
Cumulative cash flow after tax			−56	−45.9	−32.8	−16.7	−0.6	15.2	30.6	45.3	59.1	71.6	82.6

Table 5.9 Cash flow table for Example 5.4 with inflated price and cost inputs—*cont'd*

Nominal cash flows ($million)	Current value	Inflation rate (%/year)											
Year			**1**	**2**	**3**	**4**	**5**	**6**	**7**	**8**	**9**	**10**	**11**
Discount rate % per annum	16												
Present value of cash flow			−48.3	7.5	8.4	8.9	7.6	6.5	5.4	4.5	3.6	2.9	2.1
NPV	9.22												
IRR-real	9.8%												
IRR-nominal	20.8%												
Nominal cash flow after tax			−56	10.1	13.1	16.1	16.1	15.8	15.4	14.7	13.8	12.6	11
Consumer price index (CPI)		10											
Real cash flow after tax			−50.9	8.4	9.8	11	10	8.9	7.9	6.9	5.9	4.8	3.9
Present value of real cash flow			−48.3	7.5	8.4	8.9	7.6	6.5	5.4	4.5	3.6	2.8	2.1

Using Approaches 2 or 3, the NPV for the project is $9.2 million, with the DCFR 20.8% in nominal terms and 9.8% in real terms. Thus the estimated profitability is considerably less than when using Approach 1.

5.8 Effect of financing on cash flows

Section 5.4 refers briefly to the effect of financing on the effective discount rate for NPV calculations. The effect of increasing the proportion of loan to equity funds increases the profitability of the project, because of the tax-deductibility of interest paid on loan funds. In Table 5.10, Example 5.4 is extended to illustrate the effects on cash flows and NPV of the mode of financing the project. It is assumed that the interest rate on loan funds and equity funds is 16%/annum in each case and that 50% contribution of funds is from loans and 50% from equity. It is further assumed that the principal of the loan is repaid in equal amounts over the operating life of the plant (10 years), and that interest payments reduce as the principal is repaid. Cash flows are estimated on the basis of the owner's participation so that the DCFR reflects a return on equity. For the case of 100% equity funding, the real DCFR is 9.8% per annum. For the case of 50% equity funding, the real DCFR is 17.9% whilst the NPV is $16.9 million.

Thus the returns are greater with increased debt financing, but the financial risks are also higher as discussed in Section 5.4.

5.9 Sensitivity analysis and risk assessment

The economic evaluation of a project has been seen to require the estimation of cash flows over the life of the project. This estimation is based on forecasts of sales volume and selling price for products, of capital and operating costs, and of project life which are all based on expectations of markets, technology, plant performance, competitors, and the overall business economy. Because of the many influences on cash flow and the long-time horizons involved, the forecasts are necessarily imprecise; hence capital investment into such projects is a risky undertaking. For any project to proceed risks must be undertaken, but part of the economic evaluation procedure is to identify and evaluate those risks.

A distinction has sometimes been made between business risk and financial risk. The term ***business risk*** is used for any risk that might impair profitability for example risk of market failure, capital overrun, or feedstock cost increase. There is ***a financial risk*** with large investment projects that the cumulative cash flow for a project may reach such large negative proportions that the project threatens the financial security of the company making the investment. The project is most vulnerable financially at its minimum net cash position, where capital expenditure overrun, delayed plant start-up or sluggish markets may occur either individually or collectively. The company may then find that it cannot meet its interest and loan repayment commitment for the project from its other income sources. This financial risk is so serious that it deserves special consideration, especially in relation to financing the project.

Table 5.10 Cash flow table for Example 5.5 showing effects of financing.

	Current value	Inflation (%/year)	Cash flows ($million)										
Year			**1**	**2**	**3**	**4**	**5**	**6**	**7**	**8**	**9**	**10**	**11**
Fixed capital ($ million)		12	–56										
Equity contribution to fixed capital			–28										
Sales volume (t/year)				70,000	85,000	100,000	100,000	100,000	100,0.000	100,000	100,000	100,000	100,000
Selling price ($/t)	1000	9											
Sales revenue ($million)				83.2	110.1	141.2	153.9	167.7	182.8	199.3	217.2	236.7	258
Variable costs ($/t)	700	11											
Variable costs ($million)				–60.4	–81.4	–106.3	–118	–130.9	–145.3	–161.3	–179.1	–198.8	–220.6
Fixed costs (excl. interest) $million	8	10		–9.7	–10.6	–11.7	–12.9	–14.2	–15.6	–17.1	–18.9	–20.7	–22.8
Interest on loan			–4.5	–4.0	–3.6	–3.1	–2.7	–2.2	–1.8	–1.3	–0.9	–0.4	0.0
Cash flow before tax and principal repay				9.1	14.5	20.1	20.3	20.4	20.1	19.6	18.3	16.8	14.6
Investment allowance	0												
Tax depreciation rate (%)	10												

Continued

Table 5.10 Cash flow table for Example 5.5 showing effects of financing—*cont'd*

	Current value	Inflation (%/year)	Cash flows ($million)										
Year			**1**	**2**	**3**	**4**	**5**	**6**	**7**	**8**	**9**	**10**	**11**
Tax depreciation allowance				5.6	5.6	5.6	5.6	5.6	5.6	5.6	5.6	5.6	5.6
Taxable income			−4.5	3.5	8.9	14.5	14.7	14.8	14.5	14.0	12.7	11.2	9.0
Tax rate (%)	40												
Tax payments			1.8	−1.4	−3.6	−5.8	−5.9	−5.9	−5.8	−5.6	−5.1	−4.5	−3.6
Principal repayment				−2.8	−2.8	−2.8	−2.8	−2.8	−2.8	−2.8	−2.8	−2.8	−2.8
Principal owing			−28	−25.2	−22.4	−19.6	−16.8	−14	−11.2	−8.4	−5.6	−2.8	0
Cash flow after tax and principal repay			−26.2	4.9	8.1	11.5	11.6	11.7	11.5	11.2	10.4	9.5	8.2
Cumulative cash flow after tax			−26.2	−21.3	−12.1	−0.6	11.0	22.7	34.2	45.3	55.8	65.3	73.5
Discount rate % per annum	16					s							
Present value of cash flow			−22.6	3.6	5.2	6.3	5.5	4.8	4.1	3.4	2.7	2.2	1.6
NPV			16.9										
Nominal DCFR %/annum			32%										
Real DCFR %/annum			17.9										

Notes: (1) The calculated NPV and DCFR are higher than those for the case with 100% equity funding. (2) It has been assumed that the principal repayment commences in year 2, the first year of plant operation. (3) Annual interest payment has been calculated as a percentage (16%) of the principal owing in that year, after principal repayment. (4) The tax credit has been allowed on interest paid in the first year, assuming interest costs are deductible from a net positive cash flow generated by the business.

Most investment projects are unlikely to cause a company such a liquidity crisis, but the business risks must still be evaluated, since the failure of a project to provide expected returns will affect the financial performance of the company overall and its ability to carry out its future plans.

Sensitivity analysis is an important part of risk evaluation, in that it explores the effects of variation in components of cash flow on the economic viability of the project. Sensitivity analysis may be used in two ways. First, each component of cash flow may be varied by 10% (or alternatively adversely by 10%), with the remainder of inputs left unchanged, and the resultant effects on profitability determined. A table may be drawn up for a project as illustrated in Table 5.11.

The analysis in Table 5.11 identifies that profitability is most sensitive to error in product selling price, then in product sales volume, then in capital investment, and so on. The sensitivity to market factors indicated in this illustrative example is common for chemical process industry investment.

In the second method, the effects of a number of events which are considered possible are explored. These may include for example:

- 6 months delay in commencement of production
- loss of market share to a competitor, resulting in reduced sales volume and hence reduced capacity utilisation of the proposed plant
- a sharp change in the value of the exchange rate versus the US dollar

This approach is sometimes referred to as 'What if' analysis. Sensitivity analysis is readily facilitated by a computer and provided for in most evaluation packages. Sensitivity analysis enables the identification of key elements for financial success and allows a wide range of possible influences to be explored. It does not, however, attempt to quantify the probability of events or deviation from base assumptions.

A number of approaches have been developed for quantifying risks, although relatively few companies appear to have adopted those methods as part of their evaluation procedure. One approach assigns probabilities to alternative events; the outcome of each event is explored in terms of the NPV resulting from the investment. Since NPV's are additive, an 'expectation' can be estimated where

$$\begin{aligned}\text{Expectation} = {} & (\text{NPV for Event A} \times \text{Probability for Event A}) \\ & + (\text{NPV for Event B} \times \text{Probability for Event B})\end{aligned}$$

Table 5.11 Example of sensitivity analysis results for a project.

	DCFR (%)	Change in DCFR from base case
Base case	22.1	Zero
+10% capital investment	19.6	−2.5
−10% product sales volume	18.5	−3.6
−10% product selling price	16.2	−4.9
+10% variable operating costs	20.3	−1.8
+10% fixed operating costs	20.4	−1.7

An acceptable risk corresponds to a situation where the expectation (expressed as the sum of weighted NPVs) is positive.

An investment procedure can often be identified as a sequence of decision points where uncertain alternative events lead to quantifiable outcomes (expressed, e.g. as NPVs). This decision sequence can be represented as a 'decision tree'.

An example of a simplified decision tree structure for planning production capacity is shown in Fig. 8.2.

A further approach assigns probability distributions to various inputs such as raw materials price, product selling price or fixed capital cost of the plant. A Monte Carlo simulation can then be made to determine a resulting probability distribution for the NPV or DCFR of the project.

Some discussions of these approaches together with some illustrative examples are provided by Allen [6] and with a specific focus on environmental projects by Moilanen and Martin [7]. In order to estimate probability distributions or magnitudes of error for various inputs, it would be instructive to review forecasting errors made in previous projects. However, very little postproject auditing seems to be carried out by industry and even less published. One exception is a Rand Corporation Study [8] which compared operational data for 44 'pioneer' process plants with cost and performance projections. (A 'pioneer' plant is a 'first of a kind' plant for which no previous cost or operating experience is available.)

The assessment of risk, whether quantified or not, always involves a degree of subjectivity. The risk may relate to the technology adopted, the product market, the political climate, industrial relations climate at the site, or inflationary trends. Whatever the risk, it is important to recognise the human element and its potential role in estimating and in risk assessment. A project team member, committed to the success of a particular project, may be unduly optimistic and not give adequate recognition of certain risks. An excessively pessimistic view, on the other hand, will almost certainly see the project rejected.

The systematic and thorough risk assessment will often enable certain actions to be taken to reduce risks, such as further market development, a change in project definition, or selection of certain personnel for project execution. In some cases, it may point to the need for strategic changes within a company outside the domains of the project under consideration.

Most companies will seek a balanced portfolio of projects with varying levels of perceived risk and corresponding opportunity as part of its overall investment strategy.

5.10 Influence of sunk costs

From the discussions in Section 5.9, it can be appreciated that in the operational phase of a project, achieved cash flows may be less than those forecasts. In addition, capital expenditures may have exceeded the amounts sanctioned. Thus the achieved NPVs and DCFRs may well correspond to unprofitable performance. This, however, does not necessarily signal the end of the project. Whilst cash flows (after loan repayments and tax) continue to be positive, they can contribute to depreciation reserves and shareholders dividends.

The question often arises—should further capital be injected into a project which is currently nonprofitable? Another question which sometimes arises is whether a currently profitable project be abandoned in favour of a new, alternative project offering greater rewards? Each time these questions are raised, a new decision point in time is reached. In each case, the capital investment decision should be made on its own merits, and should not necessarily be bound by previous or 'sunk' investments. Estimates should be made of the net changes in cash flows, investment and operating, resulting from each new investment. Net changes in cash flow are best estimated by first establishing a base case of no investment followed by the case with the proposed investment.

5.11 Competitive analysis

Because of the many uncertainties involved in forecasting the various technical and economic parameters contributing to future cash flows, and in particular, those relating to product selling price, many companies undertake a supplementary analysis of cost competitiveness. The cost curve of operators worldwide is estimated, often with the aid of specialist consultant reports, with the aim of ensuring that the cash operating cost for the proposed project is competitive, say within the lowest 25% of those by world producers. Examples with international perspectives could examine the effects of location, exchange rate, different tax rules, and different levels of tariff protection. Other examples would focus on the technology employed, whilst others would examine aspects of project implementation regarding logistics, utility supply, integration with other plants, quality and cost of infrastructure, and so on. One specific aspect of competitive analysis with examples is discussed in Section 8.6; in this case, the production cost on an old expanded operating plant is compared with that on new plants employing the best available technology.

References

[1] The Penguin Macquarie Dictionary of Economics and Finance, Penguin Books, New York, NY, 1988.

[2] D. Brennan, Realistic discounting of cash flows, Process Econ. Int. 2 (4) (1981) 25–28, 30.

[3] United Nations, Statistical Yearbook, (2018) New York, NY.

[4] F.M. Wilkes, Inflation and capital budgeting decisions, J. Bus. Financ. 4 (3) (1972) 46–53.

[5] S.C. Leuthold, Interest rates, inflation and deflation, Financ. Anal. J. (Jan/Feb) (1981) 28–41.

[6] D.H. Allen, A Guide to the Economic Evaluation of Projects, third ed., Institution of Chemical Engineers, Rugby, UK, 1991.

[7] T. Moilanen, C. Martin, Financial Evaluation of Environmental Investments, Institution of Chemical Engineers, Rugby, UK, 1996.
[8] E.W. Merrow, et al., Understanding cost growth and performance shortfalls in pioneer process plants, Report No. R-2569-DOE, The Rand Corp., Santa Monica, CA, September 1981.

CHAPTER

Some examples in process economics

6

And suppose we solve all the problems it presents? What happens? We end up with more problems than we started with. Because that's the way problems propagate their species. A problem left to itself dries up or goes rotten. But fertilise a problem with a solution—you'll hatch out dozens.
N.F. Simpson b1919. English playwright in 'A Resounding Tinkle' (1958)

The following worked examples are extracted from previous teaching experience by the author at chemical engineering departments at RMIT, Monash and Melbourne Universities. They are presented here to illustrate some of the principles discussed in the previous chapters. The problems and solutions tend to be presented more simply than pertains in practice, though the examples are typical of industrial problems. Cost data used in the examples are for illustrative purposes only. Solutions presented are not the only justifiable solutions. Unless otherwise specified, the dollar ($) is intended as a generalised currency unit.

Example 6.1 Approximate estimate of production cost for an aluminium smelter

A two potline aluminium smelter proposed for integration with an existing alumina refinery involves a capacity of 600,000 tonnes primary aluminium per year. Major raw materials are alumina (1.93 t/t aluminium) and carbon (0.42 t/t aluminium), and the major utility is electric power (13.0 MWh/t aluminium). Other raw materials and utility costs are negligible by comparison. Fixed capital cost is estimated at 2.5 billion dollars and the smelter will employ 600 people when operating. Anticipated raw material and utility costs are

Alumina	$300/t
Carbon	$300/t
Electric power	$60/MWh

Make an approximate estimate of the production cost per tonne of primary aluminium with the smelter operating at full capacity. How does this compare with the current selling price for aluminium?

The following factors of capital cost are assumed:

$k =$	Capital recovery 5% per annum over 20 years	0.080
	+ Maintenance materials	0.025
	+ Insurance, property taxes, minor overheads	0.025
	Total	**0.13**

Process Industry Economics. https://doi.org/10.1016/B978-0-12-819466-9.00006-7

Average wage per employee is estimated at \$80,000 per year with payroll overheads estimated at 40% of employee wages.

Solution

The simplified cost model is used based on the following equation.

$$C = \sum R_i r_i + \sum E_j g_j + (Mm)/QU + (kI)/QU \quad (6.1)$$

This approach not only provides a quick estimate of production cost but also a view of the breakdown in the contributing costs. The costs are presented in the form of a simplified worksheet which can be readily developed using a spreadsheet package. The inclusion of capital recovery within capital dependent costs makes the estimate analogous to that of a floor price. The estimated production cost is similar to the reported aluminium selling price of approximately \$2100/tonne in 2018. Some factors influencing costs would be the cost of alumina which would benefit from integration with an alumina refinery, and the cost of electricity which could benefit from a favourable contract derived from a large consumption over an extended period. In developing a project for an aluminium smelter, the key cost areas and assumptions would, however, need careful scrutiny.

Plant capacity 600,000 tonnes aluminium/year
Fixed capital investment \$2.5 billion

Operating costs			\$ million/ year	\$/tonne aluminium	Fraction of total cost
Raw materials	Unit usage t/t aluminium	Unit cost \$/t			
Alumina	1.93	\$300		579	0.27
Carbon	0.42	\$300		126	0.06
Utilities	Unit usage	\$/MWh			
Electricity	13 MWh per t aluminium	60		780	0.36
Total variable cost				1485	
Personnel	Number	\$/employee			
	600	112,000	67.2	112	0.05
Fixed capital-related costs		Factor			
		0.13	325	542	0.25
Total fixed cost including capital recovery				654	
Total operating cost				2139	1.0

Example 6.2 Estimation of cooling water utility cost

A recirculated cooling water system comprises a cooling tower, cooling water recirculation pumps, and associated reticulation piping. Evaporation, entrainment, and purge losses require a make-up of freshwater of 3.5% of the circulating water. Based on the cost and performance estimates listed below, make an approximate estimate of the cost of cooling water per m^3 of recirculated cooling water. Neglect any contributions from operating or technical personnel.

Flow rate of recirculated water	$0.85\,m^3/s$
Fixed capital cost of cooling tower, recirculation pumps, and reticulation system	\$7.2 million
Electricity consumed in water recirculation, cooling tower fans recirculated cooling water	$0.6\,kWh/m^3$
Unit cost of make-up water	$\$1.2/m^3$
Unit cost of electricity	\$100/MWh

Solution

Cost of make-up water	$0.035 \times 120 = 4.2\,cents/m^3$ recirc. c.w.
Electricity cost	$10 \times 0.6 = 6.0\,cents/m^3$ recirc. c.w.
Fixed capital-related costs	
Annual cost of maintenance, insurance, property taxes, depreciation collectively estimated at 16% fixed capital	$0.16 \times 7.2 = \$1.15$ million per annum
Assuming cooling water circulates continuously for 365 days per year	
Cost per m^3 circulating water	$(1.15 \times 10^6 \times 100)/(0.85 \times 3600 \times 24 \times 365) = 4.29\,cents/m^3$
Total cost of recirculated cooling water	$= 4.2 + 6.0 + 4.3 = 14.5\,cents/m^3$

Note that the estimate excludes any operating labour or plant overhead charges, or chemical treatment or effluent treatment costs, which collectively are likely to be minor, say <10% of total costs. This is a further example of how the simplified model in Eq. (6.1) can be usefully employed to determine utility costs, and provide a breakdown in contributions to their costs. Note that in contrast to Example 6.1, the cost of depreciation rather than capital recovery has been used.

The model can be further used to explore the effects of capacity utilisation. Some industries, for example, those using agricultural products as feedstocks, are influenced by seasonal effects and require their cooling towers to be operational for limited periods. If capacity utilisation was restricted to 70% for example, fixed capital-related costs per m^3 cooling water would increase to $6.1\,cents/m^3$, bringing the total cost to $16.3\,cents/m^3$.

The estimation approach outlined could similarly be applied to steam or other utilities.

Example 6.3 Evaluation of pump options using simplified cash flow analysis

A carbon steel pump handling a corrosive slurry has been found to have an average service life of 1 year. A pump constructed in a high nickel alloy has an estimated service life of 6 years but has an installed cost of four times that of the carbon steel pump. If the economic life of the project in which the pump is to be used is 3 years from the commencement of production, and the firm is earning 15%

after tax in real terms on invested capital, which pump is preferred? The estimated salvage value of the nickel alloy pump after 3 years operation is 30% of its initial installed cost.

Possible solution

	Present value factor	Investment cash flows	
		Carbon steel pump	High nickel alloy pump
End of year 0	1.0	−X	−4X
End of year 1	0.87	−X	Zero
End of year 2	0.76	−X	Zero
End of year 3	0.66	Nil	+1.2X
Present value		−X [1+0.87+0.76]	−4X+1.2(0.66)X
		= −2.63X	= −3.21X

On the basis of the simplified cash flow analysis, the carbon steel pump should be purchased since it represents a smaller overall net present cost over the life of the project. However, considerations such as plant reliability, maintenance costs, pump removal costs, have been ignored. Tax and inflation effects have also been ignored. Assumptions regarding the cost of capital and plant operating life also influence the selection decision. A more detailed costing scrutiny would be necessary for a real case, encompassing other material options, especially where the pump duties dictate large investments.

Example 6.4 Evaluation of waste treatment options

A company invests $2.0 million over 1 year in a waste treatment process (Project A) which gives positive net annual cash flows after tax of $400,000 in each subsequent year. At the start of the third year of operation of Project A, a new alternative process (Project B) becomes available. Project B is estimated to require an investment of $1.6 million over 1 year, and to give positive annual net cash flows after tax of $800,000 in each subsequent year. It is estimated that a market exists for the products of both processes for a further 7 years after Project B becomes available. The waste must always be treated. Equipment from Project A cannot be used in Project B. Should the company continue with Project A, or should it scrap Project A and adopt Project B?

Solution outline

Decision point occurs when Project B becomes available. If Project B is built in year 1, Project A must be kept operating during that year.

	Cash flows after tax ($ million)	
Year	Project A	Project B
1	0.4	−1.2
2	0.4	0.8
3	0.4	0.8
4	0.4	0.8
5	0.4	0.8
6	0.4	0.8
7	0.4	0.8

Net present values for above cash flows ($ million)

	Project A	Project B
Discount rate % per annum		
0	2.8	3.6
5	2.31	2.72
10	1.95	2.08
15	1.66	1.59
20	1.44	1.22
DCFR % per annum	?	63

Further questions

1. How important is the choice of the initial reference point, and the original investment in Project A?
2. What is the value of the DCFR for project A and why?
3. We can see that the DCFR for Project B of 63% is very attractive. But if $i = 15\%$ for instance, it is still preferable financially to persevere with Project A. This highlights the difficulty of replacing existing process plants with plants of superior technology. The return often does not justify the capital investment required.
4. What influence might risk considerations have on the final outcome?

Example 6.5 Energy conservation project

An energy savings project has been identified in a metallurgical plant which is estimated to save $800,000 annually before tax. The project involves a waste heat boiler installation, for which the fixed capital cost estimate is $2.0 million. The project would take 1 year from now to implement. The remaining life of the metallurgical plant from now is estimated to be 6 years. The cost of capital for the project is 7% per annum, the corporate tax rate is 30%, and the allowable tax depreciation rate on plant and equipment is 10% per annum until fully depreciated.

(a) Draw up a cash flow table for the life of the project showing annual cash flows after tax.
(b) Calculate the net present value for the energy savings project and hence make a conclusion about the apparent profitability of the project.
(c) Identify three possible deviations from assumed aspects of the project which you would recommend to be considered prior to investment. Comment on their importance, and their effect on the viability of the project.

Solution

(a) Cash flow table

Year	Cash flow estimation	Present value $ million Cash flow * $(1/(1+i)^t$
1	**−$ 2.0 million**	−2* (1/1.07) = −2*0.93 = **−1.87**
2	Savings = + $0.8 million Tax depreciation allowance =0.2 Taxable income = $0.6 million Tax paid = 0.18 million **After tax cash flow = $0.62 million**	**+0.62*0.87 = 0.54**

Year	Cash flow estimation	Present value $ million Cash flow * $(1/(1+i)^t$
3	**After tax cash flow = $0.62 million**	**+0.62*0.82 = 0.51**
4	**After tax cash flow = $0.62 million**	**+0.62*0.76 = 0.47**
5	**After tax cash flow = $0.62 million**	**+0.62*0.71 = 0.44**
6	Savings = + 0.8 million Tax depreciation allowance = 1.2 Taxable income = −0.4 million Tax paid = −0.12 million **After tax cash flow = $0.92 million**	**+0.92*0.67 = 0.62**

(b) Net present value = $\sum$Present values
= − 1.87+0.54+0.51+0.47+0.44+0.62 = 0.71 million
NPV exceeds zero, hence the project appears profitable.

(c) Possible deviations could occur in

- Capital Investment—i.e. is the estimate accurate? Are there risks of significant production loss during waste heat boiler installation?
  If the capital estimate is too low and revenue losses occur during installation and commissioning, profitability will be adversely affected.
- Remaining life of the project.
  The longer the remaining life, the more profitable the project would become. If however the project life were reduced by 1 year, the project would barely break even.
- Estimate of savings.

Do the estimated savings account for any incremental labour, maintenance, and insurance costs which would reduce profitability? Does savings estimate account for any carbon tax benefits which would enhance profitability?

Example 6.6 Project evaluation—phased capital investment

A treatment plant is being designed in 2019 to remove impurities from a gas prior to sales. The sales volume is forecast to be 1 million m^3 in 2021 and to grow exponentially at a rate of 8% per annum. Gas is to be supplied until the end of 2032.

Two alternative patterns of investment are being considered. The first involves building a plant in 2020 of sufficient capacity to meet the demand at the end of the 12-year period. The second involves building two smaller plants, the first built in 2020 and sized to meet the demand at the end of 2023, and the second built in 2023 to come on stream in 2024, and to meet the balance of supply until 2032. The following economic data applies:

Working capital	= negligible
Fixed capital investment	= 20.0 $C^{0.7}$ $ million
Annual fixed operating costs	= 2.0 $C^{0.7}$ $ million
Annual variable operating costs	= 4 S $ million
Annual sales revenue	= 16 S $ million
where C = capacity of plant in 10^6 m^3/annum	
S = annual sales in 10^6 m^3/annum	
Tax depreciation allowance	= 10 % per annum
Corporate tax rate	= 30%

Tax is paid 1 year in arrears of earning income on which tax is assessed.

Solution

The cash flow patterns for the alternative schemes are outlined below

Cash flows for Scheme A, cash flows are in $million

Scheme A		0													
Year	2020	2021	2022	2023	2024	2025	2026	2027	2028	2029	2030	2031	2032	2033	**NPV**
Fixed capital	–36.2														
Sales million m^3		1	1.08	1.17	1.26	1.36	1.47	1.59	1.71	1.85	2.00	2.16	2.33		
Sales revenue		16	17.28	18.66	20.16	21.77	23.51	25.39	27.42	29.61	31.98	34.54	37.31		
Fixed costs		–3.6	–3.6	–3.6	–3.6	–3.6	–3.6	–3.6	–3.6	–3.6	–3.6	–3.6	–3.6		
Variable costs		–4	–4.32	–4.67	–5.04	–5.44	–5.88	–6.35	–6.86	–7.40	–8.00	–8.64	–9.33		
Cash flow before tax		8.38	9.34	10.38	11.50	12.71	14.02	15.43	16.95	18.60	20.37	22.29	24.36		
Tax depreciation allowance		3.62	3.62	3.62	3.62	3.62	3.62	3.62	3.62	3.62	3.62	3.62	3.62		
Taxable income		4.77	5.73	6.77	7.89	9.09	10.40	11.81	13.33	14.98	16.76	18.68	20.75		
Tax paid			–1.43	–1.72	–2.03	–2.37	–2.73	–3.12	–3.54	–4.00	–4.49	–5.03	–5.60	–6.22	
Cash flow after tax	–36.2	8.38	7.91	8.66	9.47	10.34	11.29	12.31	13.41	14.60	15.88	17.26	18.76	–6.22	
Cumulative cash flow after tax	–36.2	–27.82	–19.90	–11.24	–1.77	8.58	19.86	32.17	45.58	60.17	76.05	93.32	112.08	105.85	
Discount factor	0.9091	0.83	0.75	0.68	0.62	0.56	0.51	0.47	0.42	0.39	0.35	0.32	0.29	0.26	
Discounted cash flow	–32.91	6.93	5.95	5.92	5.88	5.84	5.79	5.74	5.69	5.63	5.57	5.50	5.44	–1.64	**35.31**

Cash flows for Scheme B, cash flows are in $million

Year	2020	2021	2022	2023	2024	2025	2026	2027	2028	2029	2030	2031	2032	2033	
Fixed capital	–22.3			–22.2											
Sales million m^3		1	1.08	1.17	1.26	1.36	1.47	1.59	1.71	1.85	2.00	2.16	2.33		
Sales revenue		16	17.28	18.66	20.16	21.77	23.51	25.39	27.42	29.61	31.98	34.54	37.31		
Fixed costs		–2.23	–2.23	–2.23	–4.45	–4.45	–4.45	–4.45	–4.45	–4.45	–4.45	–4.45	–4.45		
Variable costs		–4	–4.32	–4.67	–5.04	–5.44	–5.88	–6.35	–6.86	–7.40	–8.00	–8.64	–9.33		
Cash flow before tax		9.77	10.73	11.77	10.67	11.88	13.18	14.59	16.12	17.76	19.54	21.46	23.53		
Tax depreciation allowance		2.23	2.23	2.23	4.45	4.45	4.45	4.45	4.45	4.45	4.45	4.45	4.45		
Taxable income		7.54	8.50	9.54	6.22	7.43	8.73	10.14	11.67	13.31	15.09	17.01	19.08		
Tax paid			–2.26	–2.55	–2.86	–1.87	–2.23	–2.62	–3.04	–3.50	–3.99	–4.53	–5.10	–5.72	
Cash flow after tax	–22.3	9.77	8.47	–12.97	7.81	10.01	10.95	11.97	13.07	14.26	15.55	16.93	18.43	–5.72	
Cumulative cash flow after tax	–22.3	–12.53	–4.06	–17.03	–9.23	0.78	11.74	23.71	36.78	51.05	66.59	83.52	101.95	96.23	
Discount factor	0.9091	0.83	0.75	0.68	0.62	0.56	0.51	0.47	0.42	0.39	0.35	0.32	0.29	0.26	
Discounted cash flow	–20.27	8.07	6.36	–8.86	4.85	5.65	5.62	5.59	5.54	5.50	5.45	5.39	5.34	–1.51	**32.73**

On the basis of NPV at a discount rate of 10%/annum, Scheme A has a 12% NPV advantage over Scheme B. Scheme A also involves less capital expenditure overall than Scheme B, and for 2024–32 lower fixed costs. Scheme B, however, has some important strategic advantages. Firstly, capital investment timing can be improved to match the developing sales pattern; if for example sales growth proves to be only 5%/annum, the second plant can be deferred and may be smaller in capacity than the first plant with savings in incremental capital and fixed operating costs. The second important advantage is that the design of the second plant should benefit from the design, construction, and operational experience gained on the first plant and any improvements in available technology.

Example 6.7 Isomerisation of naphtha

SMUCE Engineering is examining the feasibility of providing a new isomerisation unit within an existing refinery in Victoria. Proposed production capacity is 1285 t/day (equivalent to 450,000 t/year) of isomerate product. The proposed plant will be constructed over 2 years in 2013 and 2014 and will commence production in 2015. The plant will operate continuously supported by a continuous shift roster of process operators. The following performance and cost data have been estimated. Costs are expressed in $Aust 2012.

Fixed capital cost	Inside battery limits: $48 million
	Outside battery limits: $9 million
Cost of initial catalyst charge	$4 million
Operating performance	
Naphtha feedstock	Consumption: 1.03 t/t product
	Unit cost: $1000/t naphtha
Hydrogen feedstock	Consumption: 0.015 t/t product
	Unit cost: $1200/t hydrogen
Catalyst make-up	$3/t product
Minor chemicals	$0.7/t product
Credit for stabiliser off-gas	$6/t product
Utilities requirements	
Electricity	40 kWh/t product
	Unit cost: $80/MWh
Steam	0.28 t/t product
	Unit cost: $15/t
Cooling water	10 m^3/t product
	Unit cost: 0.1 $/$m^3$
Process labour	3 operators per shift
	Annual salary per operator: $80,000 per year
	Payroll overheads: 40% wages

Plant overheads	80% process labour costs (inclusive of payroll overheads)
Maintenance	4% fixed capital
Insurance	1% fixed capital
Property taxes	1% fixed capital
Corporate administration, research and development, selling expenses	2% fixed capital

(a) Estimate operating costs at 100% capacity utilisation. Present costs in summary form on a detailed operating cost worksheet which includes both annual costs and costs per tonne product.
Based on the estimates for 100% capacity utilisation, estimate the operating costs for 80% capacity utilisation, expressed in both $ per year and per tonne of product.

(b) Estimate working capital requirements at 100% capacity utilisation based on an assessment of required stocks of feedstocks and finished products, and 6 weeks debtors and 6 weeks creditors. State your basis for assessing the required stocks.

(c) Estimate the annual cash flows after tax, over the life of the project. Draw up a cash flow table for the life of the project. Neglect the effects of inflation and express cash flows in $2012. Assume:
- A plant operating life of 15 years, terminating in 2029.
- Plant operates at 80% capacity utilisation in 2015, and 100% capacity utilisation in 2016, and for the remaining years of operation.
- Tax depreciation allowance of 6.7% fixed capital.
- Corporate tax rate of 30% taxable income.
- A selling price of $1150/t isomerate product, reflecting the increase in octane number achieved.

(d) (i) Assume an annual discount rate, reflecting the cost of capital, of 10% per annum. Hence calculate the NPV and IRR for the project.
(ii) Make a conclusion regarding the apparent viability of the project.
(iii) Identify what you consider to be three key factors affecting the economic viability of this project, giving your reasons.

(e) Estimate the fixed capital, working capital, and operating cost for an isomerisation unit of 900,000 tonnes/year, for the full capacity utilisation case.

Solution

Part A—Operating cost estimate

Operating cost summary sheet

Product		Isomerate	
Process route		Catalytic isomerisation	
Plant capacity	t/yr	450,000	1285.7
Capacity utilisation	%		
Fixed capital IBL	$mill	48	
Fixed capital OBL	$mill	9	
Total fixed capital	$mill	57	

Initial catalyst charge	$mill	4		
Product selling price	$/t	1150		
Production cost			**Annual cost ($million/yr)**	**Cost per tonne ($/tonne product)**
Raw materials	Unit	Unit		
	Usage	Cost		
	Unit/t	$/unit		
Naphtha t	1.03	1000	463.5	1030
Hydrogen t	0.015	1200	8.1	18
Catalyst make-up				3
Minor chemicals				0.7
Stabiliser off gas credit				–6
Total raw materials cost			470.6	1045.7
Utilities	Unit	Unit		
	Usage	Cost		
	Unit/t	$/unit		
Electricity (MWh)	0.040	80		3.2
Steam LP (t)	0.28	15		4.2
Steam IP (t)				
Steam HP (t)				
Cooling water (m3)	10	0.1		1
Process water (m3)				
Nitrogen (m3)				
Total utilities cost			3.78	8.4
Process labour	Number	Salary $/yr		
Operators/shift	3			
Shift teams	5			
Total shift operators	15	80,000	1.2	
Day operators	0			
Total process labour wages			1.2	
Payroll overheads	% wages	40	0.48	
Total process labour cost			1.68	
Plant overheads	% process labour	80	1.34	
	% fixed capital			

Maintenance	4		2.28	
Insurance	1		0.57	
Property taxes	1		0.57	
Book depreciation				
Total fixed production cost			6.44	
Total production cost		Fixed	6.44	14.3
		Variable	474.35	1054.1
		Total		1068.4
Nonmanufacturing costs	2		1.14	2.5
Corporate administration				
Research and development				
Selling expenses				
Total operating cost			481.93	1070.9
	% Total capital			
Profit	15			19
Minimum viable selling price				1089.9
Operating cost at 80% capacity utilisation		Fixed	7.58	21.1
		Variable	379.48	1054.1
		Total		1075.2

Part B **Working capital**			
Naphtha feed stocks	3 days	3.81	
Initial catalyst charge		4	
Product stocks	3 days	3.96	
Debtors	6 weeks	59.71	
Creditors	6 weeks	–54.73	
Total		16.75	
At 80% capacity utilisation		13.40	
At 100% capacity utilisation		3.35	Incremental cash flow

Part C
Cash flows are in $million

Year	2013	2014	2015	2016	2017	2018	2019	2020	2021	2022	2023	2024	2025	2026	2027	2028	2029
Fixed capital	–28.5	–28.5															
Working capital		–13.4	–3.4														16.9
Sales revenue			414	517.5	517.5	517.5	517.5	517.5	517.5	517.5	517.5	517.5	517.5	517.5	517.5	517.5	517.5
Variable costs			379.5	474.4	473.5	473.5	473.5	473.5	473.5	473.5	473.5	473.5	473.5	473.5	473.5	473.5	473.5
Fixed costs			7.58	7.58	7.58	7.58	7.58	7.58	7.58	7.58	7.58	7.58	7.58	7.58	7.58	7.58	7.58
Operating cash flow before tax			26.9	35.6	36.4	36.4	36.4	36.4	36.4	36.4	36.4	36.4	36.4	36.4	36.4	36.4	36.4
Tax depreciation allowance			3.82	3.82	3.82	3.82	3.82	3.82	3.82	3.82	3.82	3.82	3.82	3.82	3.82	3.82	3.53
Taxable income			23.12	31.75	32.60	32.60	32.60	32.60	32.60	32.60	32.60	32.60	32.60	32.60	32.60	32.60	32.89
Tax paid			6.94	9.53	9.78	9.78	9.78	9.78	9.78	9.78	9.78	9.78	9.78	9.78	9.78	9.78	9.87
Cash flow after tax	–28.5	–41.9	16.6	26.0	26.6	26.6	26.6	26.6	26.6	26.6	26.6	26.6	26.6	26.6	26.6	26.6	43.5
Cumulative cash flow	–28.5	–70.4	–53.8	–27.8	–1.1	25.5	52.2	78.8	105.4	132.1	158.7	185.4	212.0	238.6	265.3	291.9	335.4

Part D
Profitability

(i) NPV and IRR	NPV@ 0%/yr	335	$million
	NPV@10%/yr	101.6	$million
	NPV@20%/yr	28.3	$million
	NPV@30%/yr	0.2	$million
	NPV@40%/yr	−11.7	$million
	IRR	30.1%	per annum
(ii)	Project is profitable based on NPV and IRR indicators. Undiscounted payback time from start of investment is ~5.1 years		
(iii)	Some key factors affecting the economic viability of project include: • Process technology – relationship between plant design, achieved octane improvement, selling price, and fixed capital • Reliability of cost estimates – both capital and operating • Ability to achieve full capacity utilisation – dependence on plant reliability, reliability of feedstock supply		
	Other factors include the assumed plant life, possible inflation effects, corporate tax rate, and tax depreciation allowance		

Part E – scale effects

Scale effects for 900,000 t/yr		**$million**	**$/t product**
Fixed capital	$57 \times 2^{0.6}$	86.4	
Working capital	16.75×2	33.5	
Variable operating cost			1054
Labour + plant overheads	$3.02 \times 2^{0.2}$	3.86	4.3
Capital dependent costs	0.08×86.4	6.91	9.1
Total operating costs			1067.4
Assumptions			
Fixed capital	Single stream plant, 0.6 capacity exponent applies		
Working capital	Directly proportional to scale		
Variable costs	On $/tonne product basis, independent of scale		
Labour + plant overheads	Single stream plant, slight dependence on scale and 0.2 capacity exponent assumed		
	Note: other assumptions re size of manning exponent could be justified		
Capital dependent costs	%age of fixed capital remains the same, independent of scale		
Remarks	Small dependence on scale is observed because of high proportion of variable costs in total operating costs		

Example 6.8 Sulphuric acid manufacture

Cost estimation, cash flow projections, profitability evaluation, scale, capacity utilisation effects.

A company with a range of profitable investments in resources and manufacturing is considering building a plant to manufacture sulphuric acid from sulphur. Sulphuric acid would be sold to adjacent fertiliser and chemicals plants. The plant would be essentially single stream and would be a net producer of energy, resulting from heat recovery in the process. Surplus steam and electricity generated would be exported to adjacent plants. The following package of data is provided for the proposed plant. Costs are in dollars at the time of evaluation.

Plant capacity	150,000 tonnes/year of 98% w/w sulphuric acid based on a total operating time of 8000 h/year
Construction programme	Plant to be erected over a 1 year period following the year of evaluation
Operating life	Plant to be operated for 15 years following completion of the construction programme
Fixed capital cost	\$38.2 million inside battery limits
	\$9.5 million outside battery limits
Raw material	Sulphur priced at \$80/tonne used at the rate of 0.32 tonnes/tonne of acid
Electric power credit	0.2 MWh/tonne acid at \$100/MWh
Steam credit	0.6 tonnes/tonne acid at \$12/t steam
Cooling water	70 m^3/tonne acid at 12 cents/m^3 cooling water
Process labour and plant overheads	\$2.0 million/annum
Maintenance, insurance, and property taxes	\$2.8 million/annum
Corporate administration	\$0.4 million/annum
Selling price of sulphuric acid	\$100/tonne acid

(a) Based on the data package provided, determine the total cash operating costs expressed both as \$/year and as \$/tonne acid at a plant output of 150,000 tonnes of acid per year.

(b) If 4 weeks consumption of sulphur and 5 days production of sulphuric acid are required as stocks, and if the trading period is 6 weeks for debtors and 4 weeks for creditors, make an estimate of the working capital requirements when the plant is producing 150,000 tonnes annually. Materials in progress inventory can be neglected.

(c) Market forecasts predict sales of 100,000 tonnes per year in the first year of plant operation increasing to 120,000 tonnes per year in the second year and 150,000 tonnes per year in the third year of operation. Draw up a cash flow table for the life of the project, showing the annual cash flows after tax. Corporate tax is to be paid at the rate of 30% of taxable income in the year in which the income is generated. Tax depreciation allowance is 10% of fixed capital until the plant is fully depreciated.

(d) Based on the cash flows estimated in part (c) determine the net present value of the project. Ignore the effects of inflation and assume an after tax cost of capital of 10% per annum. How would you classify the economic viability of the project? Give your reasons.

(e) Based on the estimate made in part (a), estimate and compare the capital and operating costs (at full plant output) for a sulphur burning plant of the same design and having a capacity of
 (i) 100,000 tonnes/year
 (ii) 600,000 tonnes/year
 State the assumptions made.

(f) Identify some key economic aspects which affect the profitability of such a project.

Solution

(a) **Cash operating costs**

Cost category	Unit	Unit usage per t acid	Unit cost ($/unit)	Annual cost ($ million)	Cost per tonne acid
Sulphur	t	0.32	80	3.84	25.6
Electricity	MWh	−0.2	100	−3.0	−20
Steam	t	−0.6	12	−1.08	−7.2
Cooling water	m^3	70	0.12	1.34	8.4
Total variable cost				1.02	6.8
Process labour and plant overheads				2.0	7.33
Maintenance, insurance, property taxes				2.8	18.67
Corporate administration				0.4	2.67
Total fixed costs		70	0.12	5.2	34.67
Total cash operating costs				6.82	45.5

(b) **Working capital**

			Cost ($ million)	
Sulphur stocks	4 weeks	(4/52)*3.84	0.295	
Sulphuric acid stocks	5 days	(5/365)*6.82	0.093	
Debtors	**6 weeks**	**(6/52)*((0.15*50)+2.58)**	**1.96**	
Creditors	**4 weeks**	**4/52*(1.92+0.63)**	**−0.39**	
Total working capital			**1.96**	
				Increment
Working capital build-up		First year of operation	1.31	1.31
		Second year of operation	1.57	0.26
		Third year of operation	1.96	0.39

(c) **Cash flows for years 1 to 4, year 12 and year 16**

Year	**1**	**2**	**3**	**4**	**12**	**16**
Cash flows in $ million						
Fixed capital	−47.7					
Working capital	−1.31	−0.26	−0.39			1.96
Sales revenue		10	12	15	15	15
Variable costs		−0.68	−0.82	−1.02	−1.02	−1.02
Fixed costs		−5.2	−5.2	−5.2	−5.2	−5.2

Cash flow before tax[a]		4.12	5.98	8.78	8.78	8.78
Tax depreciation allowance		4.77	4.77	4.77	0	0
Taxable income		−0.65	1.21	4.01	8.78	8.78
Tax payment		−0.195	0.363	1.203	2.634	2.634
Cash flow after tax		4.315	5.617	7.577	6.146	6.146
Cash flow after tax adjusted for working capital	−49.01	4.575	6.007	7.577	6.146	8.106
Present value factor	0.909	0.826	0.751	0.683	0.319	0.218
Present value	−44.6	3.78	4.51	5.18	1.96	1.77

[a]*Excluding working capital.*

(d) Based on the cumulative cash flow, the project does not recover the capital investment until the eighth year of operation. Based on the net present value, the project does not become profitable until the 13th year of operation, achieving only $2.9 million at the end of 15 years of operation. Hence based on the assumptions made, the project is only marginally profitable

Cash flows and present values over the life of the project.

Cash flows, cumulative cash flows, present values and net present value are in $ million	Year	Cash flow	Cumulative cash flow	Discount factor ($i = 0.1$)	Present value	Cumulative present value
	1	−49	−49	0.909	−44.54	−44.54
	2	4.76	−44.24	0.826	3.93	−40.61
	3	6.01	−38.23	0.751	4.51	−36.09
	4	7.58	−30.65	0.683	5.18	−30.92
	5	7.58	−23.07	0.621	4.71	−26.21
	6	7.58	−15.49	0.565	4.28	−21.93
	7	7.58	−7.91	0.513	3.89	−18.04
	8	7.58	−0.33	0.466	3.53	−14.51
	9	7.58	7.25	0.424	3.21	−11.29
	10	7.58	14.83	0.386	2.93	−8.37
	11	7.58	22.41	0.351	2.66	−5.71
	12	6.15	28.56	0.319	1.96	−3.74
	13	6.15	34.71	0.29	1.78	−1.96
	14	6.15	40.86	0.263	1.62	−0.34
	15	6.15	47.01	0.239	1.47	1.13
	16	8.11	55.12	0.218	1.77	2.89
	NPV				**2.89**	

(e) Since sulphur burning sulphuric acid plants are essentially single stream a capacity exponent of 0.6 is assumed. Since fixed costs make-up the major contribution to total operating costs

(i) For the 100,000 tonnes per year plant, estimated capital cost
$I = (100{,}000/150{,}000)^{0.6} * 47.7 = \$37.4$ million

Variable operating cost is assumed to remain unchanged at \$6.8/tonne acid
Annual cost of process labour and plant overheads are assumed to remain essentially unchanged at \$2.0 million per year.
Maintenance, insurance, and property taxes are expected to reduce in proportion to the lower fixed capital expenditure, i.e. $(37.4/47.7)\times\$2.8 = \$2.2$ million per year.
Corporate administration costs are expected to remain unchanged at \$0.4 million per year.
Total fixed costs are estimated at $2.0+2.2+0.4=\$4.6$ million per year, or $4.6/0.1=\$46$ per tonne acid.
Total cash operating costs are thus estimated at $\$6.8+46=\$52.8$ per tonne, an increase of 16% over the base case. If we add an allowance for book depreciation based on the 15-year operating life, $(0.067\times 37.4/0.1)$, this would amount to 24.7\$/tonne bringing total operating costs to \$77.5 per tonne.

(ii) For the 600,000 tonnes per year plant
$I=(600/150)^{0.6}*47.7=\$109.6$ million
Variable operating cost is assumed to remain unchanged at \$6.8/tonne acid.
Annual cost of process labour and plant overheads are assumed to marginally increase to \$2.4 million per year.
Maintenance, insurance and property taxes are expected to increase in proportion to the higher fixed capital expenditure i.e. $(109.6/47.7)\times\$2.8=\$6.43$ million per year.
Corporate administration costs are expected to remain unchanged at \$0.4 million per year.
Total fixed costs are estimated at $2.4+6.4+0.4=\$9.2$ million per year, or $9.2/0.6=\$15.3$ per tonne acid.
Total cash operating costs are thus estimated at $6.8+9.2=\$16$/tonne acid.
An allowance for book depreciation over project life $(109.6\times 0.067)/0.6$ or \$12.2/tonne acid would bring total operating costs to \$28.2/tonne.

(f) Based on the assumptions made for the base case, the project is only marginally profitable. Closer examination of the basic cost assumptions, market evaluation, and assumed plant life would be required, as well as competitive influences from other sulphuric acid producers. Cost assumptions have not included selling expenses or royalties (alternatively research and development costs). Variable cost assumptions depend on the ability to supply surplus electricity and steam generated. Sales volumes depend on market demand, which is almost always an uncertainty. Capacity utilisation of the plant will depend on market demand as well as plant operational reliability.
Competition may occur over the life of the plant from other sulphuric acid producers.

Example 6.9 Pulp mill evaluation

A New Zealand company is making a study of a medium-capacity option for a pulp mill. The plant would be constructed in 2021 and 2022 commencing operation in 2023. The following package of economic and performance data are assembled.

Plant capacity	300,000 tonnes per annum
Fixed capital investment	900 million spent uniformly over 2 years of construction
Projected pulp sales	200,000 tonnes in 2023
	250,000 tonnes in 1994
	300,000 tonnes in 1995 and subsequent years
Operating costs	

Wood chips: consumption	$3.6\,m^3$/t pulp
Unit cost of wood chips	\$60/m^3 wood chips
Cost of chemicals	\$11/t pulp
Fuel consumption	0.11 t fuel/t pulp
Unit cost of fuel	\$450/t
Total personnel employed	360 (includes process labour, maintenance labour, technical, clerical, and administrative staff)
Average salary per employee	\$80,000 per annum
Selling price of pulp	\$1200/tonne pulp ex-mill
Credit for exported electric power	\$20/tonne pulp

(a) Estimate the total annual operating cost based on 100% capacity utilisation
Expressed as annual cost and cost per tonne of pulp product.

(b) Estimate the working capital at 100% capacity utilisation assuming:
6 weeks product stock
6 weeks debtors
4 weeks of fuel oil stock
4 weeks stock of wood chips
6 weeks stock of chemicals
4 weeks creditors
Salaries paid 2 weeks in arrears

(c) Construct a cash flow table detailing the annual cash flows after tax for the life of the project. Ignore inflation and financing charges. Assume
Plant operating life = 15 years
Annual tax depreciation allowance 6.7 % until fully depreciated
Corporate tax rate: 30%
Tax paid in year income is generated.

(d) Based on cash flows after tax, estimate the payback time and net present value for the project based on a discount rate of 10% per annum. Comment on the apparent economic viability of the project.

Solution

To be provided by the reader.

Example 6.10 Market evaluation

For one or more of the products nominated, explore

- dominant raw material and manufacturing process; main cost drivers in manufacture (e.g. feedstock, utilities, capital)
- product end uses; product quality requirements; means of supply (e.g. packaged, bulk, pipeline)
- current demand, demand growth pattern, in major economies; current (or recent) selling price
- potential for investment into new or expanded capacity in a country of your choice
- a recommended manufacturing site, supported by reasons for selection

Products can be selected for example from:

Ammonia, oxygen, hydrogen, caustic soda, nitric acid, acetone, phosphoric acid, titanium dioxide, methanol, fuel ethanol, or other products of interest.

Solution

To be provided by the reader.

CHAPTER

Process technology evolution and adoption

7

Evolution.... is—a change from an indefinite incoherent homogeneity, to a definite coherent heterogeneity.

Herbert Spencer, English Philosopher (1820–1903) from 'First Principles' (1862)

Chapter outline

Process Industry Economics. https://doi.org/10.1016/B978-0-12-819466-9.00007-9

7.1 Introduction

Chemical engineers are required to evaluate process technologies for the conversion of raw materials to products in a number of different contexts:

- the research and development phase of an emerging process
- the selection of a process from commercially established alternatives
- decision-making in the design of a process plant for a project
- the comparison of a process owned or operated by a company with those owned or operated by competitors
- product evaluation, where life cycle assessment criteria demand consideration of the manufacturing process chain

When making these evaluations a number of criteria are important, including

- product quality assurance
- capability to handle variations in feedstock quality
- capital cost per unit of production capacity
- raw materials consumption per tonne of the product
- energy and utility consumptions per tonne of the product
- operating cost per unit of production
- health and safety standards, especially in relation to toxicity, fire and explosion hazards
- environmental impact, especially in relation to emissions and wastes generated
- operational reliability
- overall sustainability

The relative importance of evaluation criteria in an assessment will depend on the type of process under evaluation and the context of the evaluation. The information upon which the assessment is made may be of varying detail and reliability and will usually come from a number of sources. These sources include

- Public domain literature, including encyclopaedias of chemical technology, books, journals, and government reports.
- Commercial service reports from consultants. Well-developed flowsheets and equipment specifications may be included and the supporting patent and technical literature reviewed. Companies can pay for this service on an initial or ongoing basis; generally, the costs are large enough to preclude small companies or individuals from participating.
- Private company sources such as reports, files, and manuals, supported by professional advice from current employees built on experience, as well as advice from companies with whom business relationships have been established. Specific reports may also be commissioned from contractors and process licensors under confidentiality agreements.

In order to make a balanced appraisal of process technology, the information sources collectively need to address the range of assessment criteria. An appreciation of the potential limitations and strengths of the various information sources, as well as possible biases by authors with vested interests, is an important element in the appraisal. A critical approach is thus necessary. The greater the range of perspectives in both the evaluation criteria and public/private domain intelligence, the more balanced and realistic the appraisal is likely to be.

An important dimension in the evaluation of a given process technology and competing technologies, is that of time. The potential for a technology to improve, and achieve an ultimate performance level will shape the business planning and research priorities of the more able process industry companies. It is also important to realise that the user of a particular process technology may not be employing the best available features of the technology. Much depends on the technical capability of the user, the point at which the technology was adopted (usually embodied in a process plant), and the extent to which subsequent improvements in the technology have been adopted and implemented.

7.2 Time dependence of process technology performance

Products and processes have lives during which the associated technology is continually developed and improved. The lives may be relatively short as with certain pharmaceutical products or quite long—for example cryogenic oxygen production and aluminium smelting based on the Hall–Heroult process are each more than 100 years old. Process plants embodying the contributing technologies of product, process, equipment, and instrumentation can become noncompetitive during their operating lives, whilst more advanced technology is developing or being implemented. It is important as a user of process technology to understand its level of maturity as well as that of technologies used by competitors. It is also important to recognise the potential of both existing and emerging technologies. In a research and development environment, it is necessary to weigh up the merits of investing further time and effort on improving an existing technology, in preference to developing an emerging technology, which whilst currently inferior, may have the potential to displace the existing technology in time.

Thus it is necessary to understand the ***dynamics of technologies***. It is useful in this context to use simple models to quantify and forecast technological change. The two models most commonly used are the sigmoid (S-shaped) curve model and the exponential model. The sigmoid curve model, based on analogy to biological growth, implies that growth is initially slow but then increases steadily for a period, after which it slows. The exponential curve can be considered as a component of a longer duration sigmoid curve, thus representing development before the rate of change begins to decrease under the influence of boundary constraints. In many cases, however, exponential development has been arrested abruptly with the onset of a plateau in performance.

Two better known sigmoid curves are the logistic curve defined by

$$Y = Y_u/[1 + A\exp(-Bt)] \tag{7.1}$$

and the Gompertz curve defined by

$$\ln Y = \ln Y_u - A\exp(-Bt) \tag{7.2}$$

where

Y = performance at time t
Y_u = ultimate performance
A, B = parameters

The point of inflexion for the logistic curve occurs at $Y = Y_u/2$ and the curve is skew-symmetric about this point. The point of inflexion for the Gompertz curve occurs at $Y = Y_u/\exp(1)$. Both curves are asymptotic to $Y = 0$ when $t \rightarrow -\infty$, and to $Y = Y_u$ when $t \rightarrow +\infty$.

Apart from the use of sigmoid curves for the quantitative modelling and forecasting of technological progress, sigmoid curves are a useful qualitative basis for representing the life (or life cycle) of a process technology. Three phases can be identified, as shown in Fig. 7.1:

- an initiation phase in which the technology is developed to a point where manufacturing commences
- a growth phase in which the process and plant are improved with a reduction in cost and increase in the scale of manufacture
- a maturity phase in which the rate of cost reduction and scale development declines markedly

Given that both demand for further improvement and appropriate ingenuity both exist beyond the point of maturity, there is an opportunity and probability for superior technology to develop in place of existing technology. Fig. 7.2 shows a generalised case where Technology B overtakes the performance of Technology A. The two curves considered together suggest an exponential trend in overall capability with time, though it can be anticipated that this capability will eventually become S-shaped. This

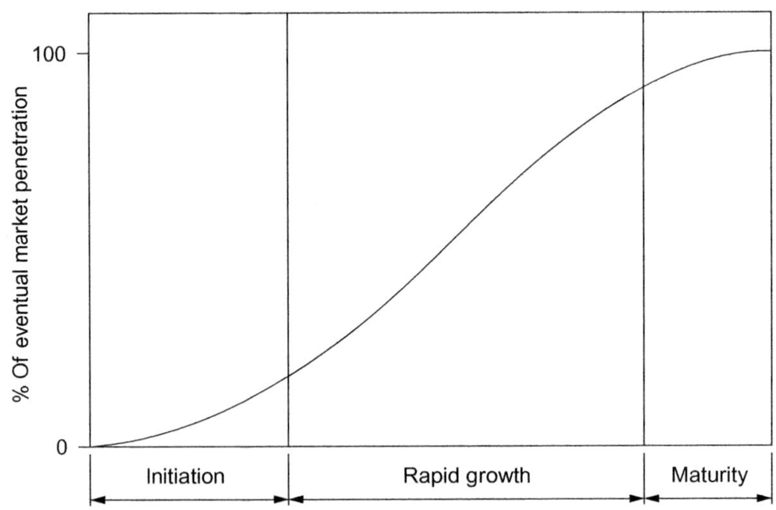

FIG. 7.1

Technology life cycle.

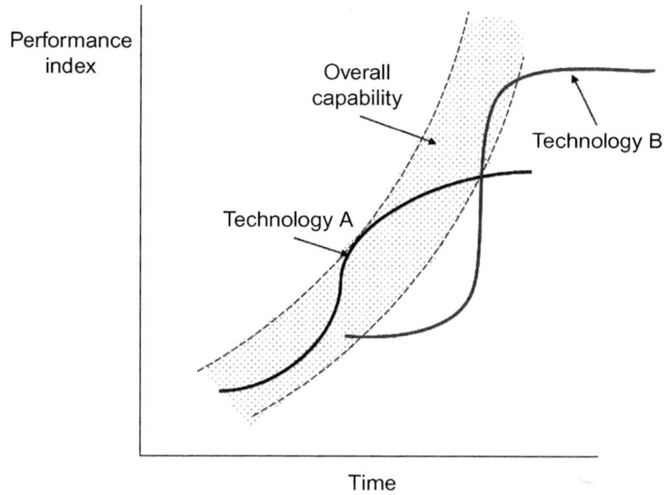

FIG. 7.2

Linear plot of technology performance versus time showing Technology B overtaking technology A.

overall capability has been described by Ayres [1] as the *'envelope' of technological change*. Examples in the literature of early technology systems include

- efficiency development in energy conversion systems derived from external fuel combustion [2]
- efficiency development in incandescent and fluorescent lighting [3]

There are numerous cases where process technologies for manufacturing specific chemicals have been progressively overtaken by superior technologies. Just two examples include

- Sulphuric acid manufacture where the lead chamber process used in the 19th century was overtaken by the contact process in the 20th century, with double absorption systems subsequently overtaking single absorption systems within the contact process.
- Chlorine and caustic soda manufacture where there were progressive improvements in mercury cells in the first half of the 20th century, followed by replacement of graphite anodes with titanium anodes in the 1960s, followed by progressive replacement of mercury cells with membrane cells from the early 1980s to 2019.

7.3 Initiation phase of technology

The initiation phase of a process technology can be lengthy. There is typically a timescale of up to 10 years or longer from initial invention or discovery to successful commercial production. The technology may be centred around a new product or a new process for making an established product. Need for a new process to make an established product may be driven by cost, feedstock availability, environmental, or safety considerations. Research on new technology is characteristically science intensive but with the need to develop the basic ideas into a commercial process making a commercial product. Thus research and development is multidisciplinary with the interaction between scientists (mainly chemists, physicists, and biologists in the case of process technologies) and engineers (mainly chemical) but also involving market and commercial appraisal. If commercial production is to result, capital costs, operating costs, and sales revenue must be forecast with sufficient certainty to justify ongoing research costs, and ultimately the capital investment for a manufacturing plant.

Because of the novelty of the emerging technology, considerable uncertainty exists in the definition of the process and in the cost and performance of the associated plant. As research and development proceed, the uncertainty is reduced but is always greater than for an established process where there is a benefit of design, construction, and operating experience on previous plants. Notwithstanding some uncertainty and the desirability of further research, there are frequently conflicting market pressures favouring early manufacture. A balance is sought between technical definition and market opportunity, in developing a manufacturing strategy, which inevitably involves a calculated risk.

From the initiation of commercial production, a further 5–10 years may elapse before production is significantly profitable. There may be slowness in the market place to adopt a new product. The slowness may be due to attitudinal inertia, but there are frequently technical problems and investment barriers associated with making a change to a new product. Thus market growth is initially restricted. The production capacity of the first plant built is frequently small, reflecting market volume or a desire to minimise the amount of capital risked in initial manufacture. Technical difficulties experienced in plant operations are also likely in the early years.

The risks associated with developing and implementing new process technologies are thus considerable. Technical, economic, and market uncertainties are present and are exacerbated by the long-time horizons involved. Safety and environmental risks also demand scrutiny. There is the ever-present risk of a competitor developing a new process or product. Costs of research and development are also significant; costs are incurred in laboratory equipment, buildings, and reagents, but mainly in personnel engaged in the work. Additional personnel are required in technical, administration, marketing, commercial, and legal functions to support the personnel directly involved in research and development.

Companies generally allocate budgets for research and development in specific areas of business. The extent of the budgets reflects the perceived life cycles for products and processes in those business areas. Thus budgets for R&D might amount to 10% of sales for pharmaceuticals and agricultural chemicals, 5% of sales for dyestuffs and polymers, and 2% for bulk and basic chemicals [4].

Despite the associated risks and costs of R&D, rewards can be very significant. Survival in existing business areas can be ensured, new areas of business can be opened up, and there are opportunities to license aspects of the developed technologies. The penalties for failure to innovate, on the other hand, can be severe, with loss of commercial initiative and in extreme cases serious business failure.

7.4 Growth phase of technology

During the growth phase of process technology, there is growth in production and sales of the associated product. Concurrent with this market growth, there is frequent growth in the production capacity of plants to manufacture that product. For many major chemicals in the past, the growth rate in maximum plant capacity has exceeded the market growth rate for the product.

As the technology is developed and improved, progressive cost reduction maintains market growth; market growth in turn encourages larger production plants, which achieve economies of scale; economies of scale enable further cost reduction sustaining the pattern of behaviour.

Economies of scale are discussed further in this chapter in Section 7.6. Strictly, economies of scale apply to cost reduction per tonne of product in a static (or constant) technological environment, as a result of increased scale or size of the operation. Economies of scale apply to both capital and operating costs. Over time, successive plants employing particular process technologies are typically larger in physical size but also achieve higher production capacity and performance improvement. These benefits are derived from technological innovations achieved through research and development, as well as the cumulative benefits of experience in design, construction, and operation.

The high market growth rates characteristic of many chemicals in the mid to late 1900s encouraged progressively larger capacity plants to be built. Simmonds [5] presented data showing growth in the maximum capacity of plants or equipment in the US industries and provided examples of technological innovation influencing such growth. Brennan [6] has also provided data on growth in the maximum available capacities for aluminium smelting, chlorine/caustic soda, and ethylene plants.

In examining capacity development in plants and equipment items, there is an important distinction between mere physical size increase and intensification effects. Scaling up of plants and processes, even with little technological change, brings its own set of technical problems. Ball and Pearson [7] examined some technical, economic, and managerial restraints in the development of plant scale over time. Criteria of similarity in mechanical, thermal, and chemical aspects have conflicted with many chemical and metallurgical plants. Constraints to scale development may be concentrated at a particular size barrier providing a number of obstacles to be overcome; once overcome however, a significant step increase in size may result. Alternatively, constraints may be dispersed at various size barriers, leading to a more gradual development of scale as problems are overcome.

Pratten [8] emphasised the continuous change in technical progress, which gives rise to different 'scale curves' for succeeding vintages of a plant (Fig. 7.3). These curves are often kinked slightly above the maximum capacity for which prior experience has been obtained, reflecting the technical and economic uncertainties of the initial transition to a larger-scale plant. Through time, the kink in scale curve is likely to be pushed to higher levels of capacity. When the ultimate limit to capacity is reached, the kink may eventually be eliminated.

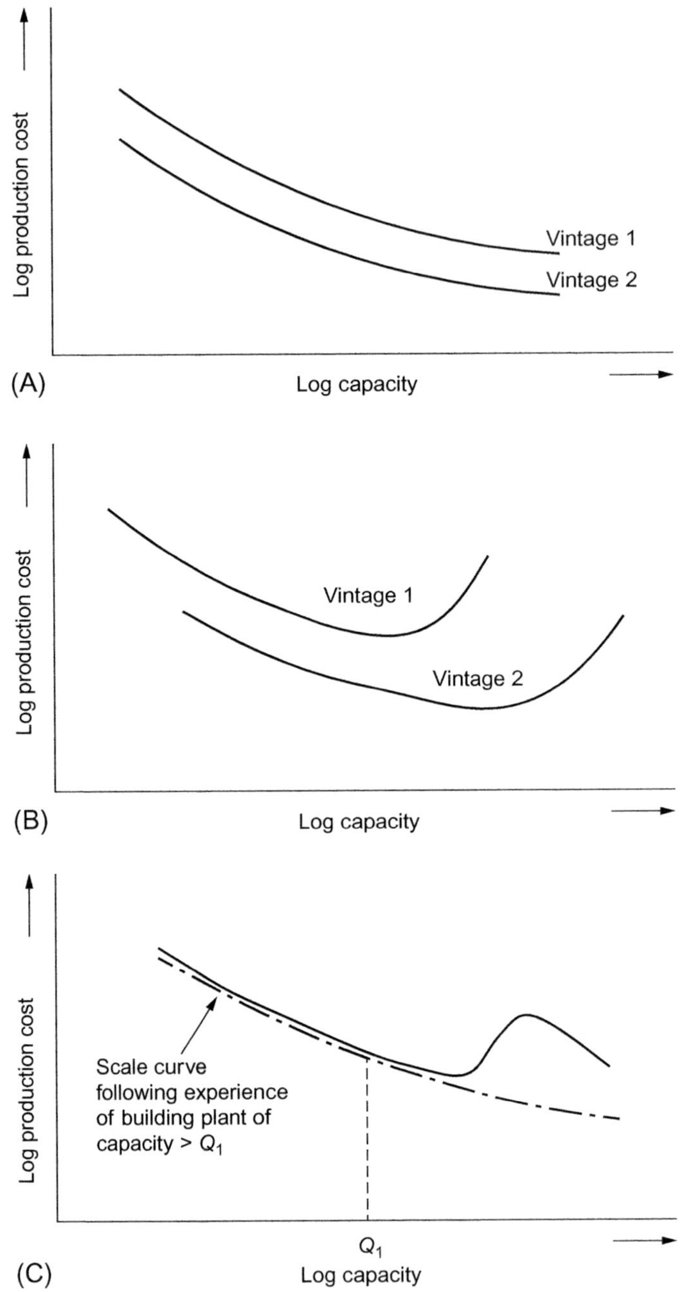

FIG. 7.3

Possible scale curves for different vintage plants: (A) unkinked; (B) kinked; and (C) alternative type of kinked curve.

7.5 Intensification

Intensification has been viewed as an important objective in process and equipment design. In the case of equipment design, 'intensification' means an increase in capacity for an equipment item of a given size, or alternatively a decrease in the physical size of equipment for a given process duty. This offers the prospect of reduced fixed capital costs including a reduction in installation labour and a reduced inventory of process material. The reduced inventory brings a consequent reduction in working capital and importantly, in many cases, in process hazards.

Intensification is commonly associated with a change in technology. Opportunities for intensification in reactors for example include changes in the catalyst, particle size or shape of feed materials, reactor design, materials of construction, and control methods. Intensification has been occurring steadily in process equipment over the decades. In packed towers used for gas absorption and distillation, improvements in packings from Raschig rings through improved rings and saddles to structured mesh packings, along with improvements in liquid and gas distributors have resulted in a reduction in tower size for a given process duty. The use of finned and ribbed surfaces instead of plain tubes has enabled more compact heat exchangers. The possibility of achieving more dramatic reductions in equipment size using centrifugal force in mass transfer and heat transfer equipment design has been explored; the Higee distillation column is one well-publicised example [9].

Process intensification (as distinct from equipment intensification) is derived from several influences. Yield and selectivity improvements in reactors increase reactor productivity not only through increased output of product per reactor volume and per input of raw material, but by unloading capacity requirements in downstream separation and purification plant. For a new plant, this offers reductions in capital and energy costs in downstream plant and reduced recycling of unreacted feedstocks. In an operating plant, capacity margins created in downstream equipment and in recycle flows offer an opportunity for economic expansion of plant production capacity. Other means of process intensification include reducing the number of process steps, reducing batch times for a batch process, or converting a batch process to a continuous process.

7.6 Economies of scale

Engineers are required to examine capacity options in plant and equipment design, at various phases of projects. This applies to an early strategic decision regarding the optimal match of plant capacity with forecast market growth (see Chapter 8 for discussion). It also applies to the detailed design of a section of the plant, where generous capacity in one equipment item can reduce the required capacity of another equipment item, in providing overall system capacity; a simple example is sizing a pipeline in relation to compressor duty in a gas transport system.

For such decisions, the engineer relies on empirically developed relationships assembled from previous cost data, or from current estimates, which relate elements of cost to capacity. Those elements which are commonly dependent on capacity in a static technological and unit cost framework are fixed capital investment, those operating costs such as maintenance, insurance and capital charges which are a function of capital investment, and operating personnel levels. The relationships are generally expressed as

$$I = k_i Q^b \tag{7.3}$$

$$M = k_m Q^a \tag{7.4}$$

where

I = fixed capital investment
M = number of persons employed
Q = plant capacity
a, b = exponents
k_i, k_m = parameters

Eq. (7.3) is used for process plants and also for both purchased and installed equipment. Eq. (7.4) is used for various categories of employees such as process labour, maintenance labour, technical, and administrative personnel.

Economies of scale apply where the values of exponents 'a' and 'b' are <1. Values of 'b' for both entire plants and equipment are published in the engineering and economics literature. For equipment of wide application (vessels, columns, heat exchangers, pumps, and compressors) data are more plentiful than equipment of specialised application, especially where the item is an integral part of a technology (such as an ethylene cracking furnace). Values of 'a' for personnel categories are less plentiful than values of 'b' for capital investment into plant and equipment. Limited published data indicate values of 'a' can differ for various categories of personnel (see Chapter 4 for further discussion).

Despite the wide use of the relationships (7.3), (7.4) for entire process plants, it is usually difficult to obtain detailed data to confirm their validity and even where such data are available, to obtain details of the basis for that data. The exponents 'b' and 'a' depend on the flowsheet structure of a plant, as discussed in Chapter 3.

In considering the effects of scale on production costs, raw material consumption and (generally) energy consumption per tonne of the product are independent of scale. An exception to the energy case exists for processes operating at extremes of temperature as discussed in Chapter 4. Personnel and fixed capital-related costs per tonne of the product are almost always scale dependent. The magnitude of a production cost penalty for a small-scale plant thus depends on its capacity relative to the maximum available plant capacity for the technology concerned, the flowsheet structure of the plant, and the relative proportion of fixed and variable costs in total production cost. The quantification of scale effects in the capital and operating costs and corresponding data are discussed in greater detail in the chapters on capital and operating cost estimation.

In the process industries, in particular, several separate process plants may be required to convert raw material to a finished product. In the case of PVC pipe for example ethylene (and chlorine), vinyl chloride monomer, polyvinyl chloride, and PVC pipe fabrication plants combine to form the processing chain. Production cost penalties from a sequence of small-scale plants are thus cumulative, influencing the costs of intermediates through to the cost of the end product.

Despite the generalisation that large-scale plants are more cost-effective than small-scale plants, a wide distribution of plant sizes is evident throughout the world for most technologies. For example, recent data for membrane cell chlor-alkali plants in Europe reported plant capacities ranging from approximately 20–650 ktonnes per year [10].

There are important cases where small capacity plants are justified, for example:

- when limited availability of cheap feedstock can make small plants competitive with large plants
- where the cost of transport relative to the product selling price is high
- where products are hazardous (for example liquefied chlorine gas), transportation may be prohibited, enabling small-scale manufacture
- where technological innovation can achieve a significant reduction in capital cost

The evaluation of scale economies in process plants has been reviewed in some detail by Brennan [11].

7.6.1 Scale-up or scale-out

Increasing plant scale for a given process technology implies increasing the production capacity at each stage of the process. This can be achieved by

- increasing the scale of process equipment (a scale-up approach) or
- increasing the number of equipment items (a scale-out approach)

The scale-out approach has been used in established process technologies where practical or economic limits in the maximum size of the equipment has been reached. Examples include reactors within

- aluminium smelters
- chlor-alkali plants
- seawater desalination plants
- ethylene plants

The scale-out approach is also adopted in renewable energy projects involving solar panels or wind turbines.

Some considerations in estimating the installed costs of duplicate items are listed in Table 7.1 for the cases of process vessels (including reactors) and for shell and tube heat exchangers. These indicate that economies of scale for multiple equipment items are more favourable for installed costs than for purchased costs. Table 7.2 lists some estimates by the author of purchased and installed costs of single and multiple items for shell and tube heat exchangers; whilst the estimates are only approximate,

Table 7.1 Some installed equipment cost considerations for duplicate vessels or heat exchangers.

Cost category	Consideration
Equipment installation	Space requirements are generally less for two items than twice the space for a single item
Piping	Potential commonality for main feed supply, product, and relief system piping
Instrumentation	Essentially duplicated
Electrical	Minimal effect
Civils	Potential for shared foundations. Some commonality for site preparation
Structural steel	Potential for shared platforms and access ways where applicable
Insulation/ fireproofing	Potential for savings through fireproofing of common supports. Potential for labour economies in installation of fireproofing and insulation
Painting	Essentially duplicated
Design	Design for two items similar to that for one item
Procurement	Similar cost for two items to that for one item. Potential small discount for second item
Project management	Some potential savings

Table 7.2 Estimates of purchased and installed costs of shell and tube heat exchangers of different size ($Australia 2018).

Item	Equipment description	Capacity	Purchase equipment cost (million $)	Total installed cost including overheads (million $)	Ratio of installed cost to purchased cost
E101	Heat exchanger—base case	1 shell @ 300 m^2	0.160	0.514	3.2
E102	Heat exchanger—double capacity	1 shell @ 600 m^2	0.263	0.756	2.9
E103	Heat exchanger—double capacity	2 shells @ 300 m^2	0.320	0.826	2.6
E104	Heat exchanger—triple capacity	3 shells @ 300 m^2	0.481	1.060	2.2

Table 7.3 Summary of implied scale exponents for costs of scaled-up and scaled-out heat exchangers (based on estimates in Table 7.2).

Item	Base unit	Scaling strategy	Scale exponent *b* Purchased cost basis	Scale exponent *b* Installed cost basis
E102	E101	Scale-up; base × 2	0.71	0.56
E103	E101	Scale-out; multiple × 2	1.0	0.68
E104	E101	Scale-out; multiple × 3	1.0	0.66

the derived scale exponents shown in Table 7.3 demonstrate that limited scale economies are still achievable for the scale-out approach.

There are contrasting effects of scale-up and scale-out in relation to safety and environmental benefits. For multiple items, the process inventory per equipment item is smaller than for a single stream plant; the potential scale of damage resulting from the release of process material from a smaller equipment item is reduced. However, the resulting impact must still be carefully considered, especially where domino effects are possible in conjunction with other equipment in the suite of multiples. The total number of potential emission points from flanges, seals, and relief valves also increases in multiple streamed plants leading to an increased potential for

- fugitive emissions, which may contribute to environmental problems
- release of toxic or flammable materials contributing to safety hazards

Both these risks may also contribute to the increased cost of monitoring and surveillance.

There are some potential advantages in overall plant reliability with multiple rather than single equipment items. If a plant relies on a single reactor and that reactor fails, the plant must be shut down for maintenance. If multiple reactors are used and one reactor fails, the plant can still be operated at reduced output. The provision of duplicate operating pumps or compressors is often used to ensure increased overall plant reliability.

7.7 Maturity

The mature phase of technology may be characterised by a pronounced slowing in market growth for the associated product, a pronounced slowing in capacity growth for plant and equipment, or a pronounced slowing in the rate of technological change. Often these signs of maturity are coincident, and the prime cause of retardation in progress may be difficult to identify.

Factors contributing to this limit include

- reduced gains in capital efficiency with increased capacity
- difficulties in raising the large capital required for larger plants
- possible complexities in project management for larger plants during design, construction, and commissioning phases leading to delays and increased costs
- susceptibility to large financial losses in the event of reduced capacity utilisation, caused either by the market or plant reliability limitations
- physical limitations in available sizes of essential equipment such as compressors, reactors, and furnaces
- larger inventories of hazardous materials

The apparent maturity for many products and technologies of the chemical industry has spurred companies in the major economies worldwide to make strenuous efforts in research devoted to new products and technologies. There has also been increased licensing of process technologies and provision of technical and management expertise by companies in the United States and Europe to countries at an earlier phase of industrial development such as Korea, Taiwan, and China. These trends serve to highlight an important aspect of technology evolution in that a given process technology may have quite distinct life cycle curves in different countries or cultures. The adoption of ethylene and ethylene derivatives technologies for example occurred some 10 years later in Japan than in the United States, leading to more rapid production growth rates at that point in Japan (35%/annum) compared with those in the United States (11%/annum).

7.8 Modelling performance improvement

Evidence for performance improvement in the process industries has frequently been presented as correlations of the real selling price (or real production cost) with cumulative production for given products, see Fig. 7.4 for example. The correlation is expressed as

$$S = dX^q \qquad (7.5)$$

where

S = real selling price ($/tonne product)
X = cumulative production (tonnes product)
d = proportionality factor
q = an exponent having a negative value

Plots of correlations of Eq. (7.5) have been variously described as 'experience curves', 'performance curves', 'improvement curves', and also 'learning curves'. Some authors have distinguished 'experience curves' incorporating the effects of scale, innovation, and learning, from 'learning curves' describing the increased

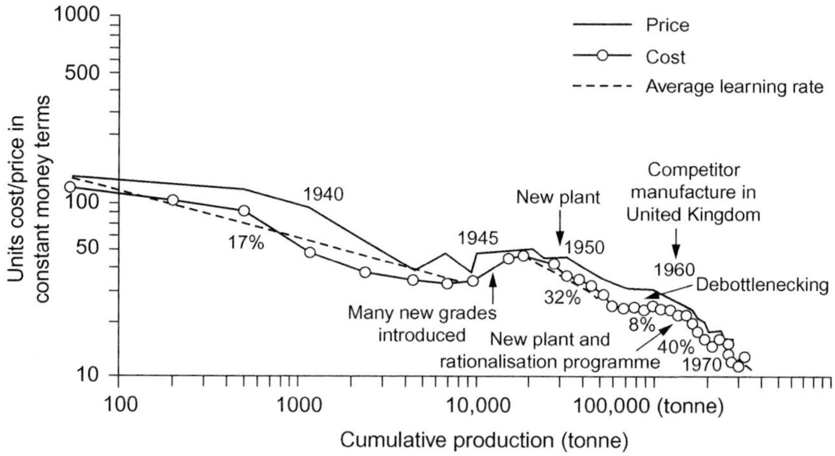

FIG. 7.4

Malpas [12] Experience curve for a specific product showing advantage from new production facilities.

efficiency of labour during repetitive production runs. Other authors make no such distinction and no further distinction in terminology will be made here.

The concept of the learning curve originated in the aircraft industry, where it was observed that the direct labour required to produce an airframe could be related to the cumulative production of airframes. Labour-intensive operations were observed to have more rapid performance improvement than machine intensive operations. Hirschmann [13] argued that the same function could be used to model performance improvement in the petroleum refining industry over extended periods, both in fixed capital costs of new refinery units built at different times and in the capacity growth of individual refinery units over their operating lives. The work of the Boston Consulting Group (BCG) in the late 1960s [14] gave further weight to the idea of the general applicability of the learning curve. The BCG observed a characteristic reduction of 20%–30% in real costs with each doubling of cumulative production experience for a diverse range of industries and products. For a 20% reduction in real costs with each doubling of cumulative production, the value of q in Eq. (7.5) would be -0.32.

Taylor and Craven [15] have explored learning curves for a range of chemical products. Taylor and Craven found that the characteristic decline in selling price occurred only for products demonstrating a high growth rate in demand. Whilst some industrialists have argued strongly for the use of learning curves in various aspects of business planning, there is still disagreement about the fundamentals and applicability of learning curves.

Brennan [6] has outlined some limitations of the learning curve theory in the light of studies of performance improvement in new plants and operating plants. Changes in selling price or production cost may reflect changes in performance or alternatively changes in unit costs of production. Following the crude oil price rise of 1973, many chemicals showed a jump in real selling price attributable to the increased cost of feedstocks and energy. One such product was ethylene. During the 1960s, reductions in the real selling price of ethylene approximated 7%/annum, but this rate of price reduction was partly enabled by a reduction in real unit costs of feedstocks and energy. An important conceptual limitation of the cumulative production function in the learning context is that it fails to account for the impact of new developments derived from technologies outside the immediate technology of the product or process concerned. For example, major developments in caustic soda/chlorine technology have been achieved through new and improved materials (especially in relation to metal anodes) and in membrane separation technology. These developments, whilst influenced by the need for application in a mature chlorine industry, are heavily influenced by technological development outside the caustic/chlorine industry. A further limitation of the relationship expressed in Eq. (7.5) is that it fails to reflect adequately the approaches to limits evidenced in a number of performance aspects for both new and operating plants.

A preferred approach for modelling performance improvement is the modelling of key aspects of performance with time. Such key aspects include

- production capacity of an entire process plant
- production capacity of individual equipment items
- raw material consumption/tonne of the product
- energy consumption/tonne of the product
- number of persons employed/annual tonne of the product
- fixed capital cost (real basis)/annual tonne of the product

This approach accommodates the diversity of progress observed for different situations and enables an insight into the key contributor to progress and possible trade-offs in one performance area versus another.

Table 7.4 summarises some rates of change embodied in new plants of best available technology and scale. Plant capacity growth and productivity improvement in raw materials, energy, persons employed, and fixed capital for four process technologies are represented. The data on aluminium smelting, chlorine/caustic soda, ethylene, and oxygen production were derived in studies by Brennan [16] and Decleva [17] whilst the data on catalytic cracking was extracted by Brennan from an earlier study by Enos [18]. The table includes estimates of production cost productivity improvement made using Eq. (4.1) and estimates of typical unit costs.

Rates of progress are expressed as equivalent exponential changes over time. Exponential change is a good approximation in some cases, for example growth in maximum plant (and equipment) scale and improvement in personnel productivity. In other cases, exponential change is less realistic but an exponential model allows a single number for the rate of change. A cyclic pattern has been observed in real fixed capital cost/annual tonne of product for ethylene plants over an extended period. The cyclic pattern reflects various phases of development in ethylene plant technology. Initially, rapid scale development improved plant integration and a

Table 7.4 Performance improvement over time in new plants employing best available technology and scale, expressed as annual rate of change.

Technology	Aluminium smelting	Chlorine caustic soda Hg cell	Ethylene from naphtha	PSA oxygen	Catalytic cracking of petroleum feedstocks
Maximum plant capacity (%/yr)	1900–80: 6.3	1940–75: 8.7 1975–85: 0.0	1950–70: 13.0 1970–85: 1.5	1971–76: 33.3 1976–79: 31.7 1976–89: 4.6	1913–53: 15.0
Raw materials consumption (%/yr)	1930–80 Alumina 0.0 Carbon 0.74	1940–80 Salt 0.0	Naphtha 1950–70: 1.9 1970–85: 1.2		1914–55: 3.3 (per 100 miles transportation)
Energy consumption (%/yr)	1945–60: 1.7 1960–82: 0.6	DC electricity 1900–75: 0.0 1975–80: 3.6	1950–60: 1.0 1960–85: 4.6	1973–83: 2.3 1983–86: 8.6	1914–55: 6.3 (per 100 miles transportation)
Personnel productivity (%/yr)	1930–80: 5.3	1940–75: 8.0 1975–80: 2.0	1950–69: 9.0 1969–85: 1.4	1986–88: 3.8	1914–55: 12.1 (per 100 miles transportation)
Fixed capital productivity (%/yr)	1940–70: 0.0 1970–80: −2.2	1940–72: 5.6 1972–80: −7.2	1950–66: 7.5 1966–85: −7.4	1973–86: 8.0 1986–89: 2.2	1914–55: 6.7 (per 100 miles transportation)
Total production cost productivity (%/yr)	1930–40: 2.7 1940–50: 2.3 1950–60: 1.8 1960–70: 1.3 1970–80: 0.2	1940–50: 3.3 1950–60: 3.0 1960–70: 3.2 1970–80: 0.0	1950–60: 2.5 1960–70: 2.2 1970–80: 0.9	1970–75: 21.0 1975–80: 7.1 1980–85: 6.6 1985–90: 3.1	1914–55: 3.7 (based on 1955 cost structure)

general drive to minimise capital costs improved capital productivity. From the late 1960s to the early 1980s, a thrust to improve reliability, safety, and environmental standards, a plateauing of growth in plant scale, and increased expenditure to reduce feedstock and energy consumption and maximise by-product recovery all contributed to decreasing capital productivity. Clearly, some compensating benefits in operating cost productivity accrued.

The annual chlor-alkali industry review published by Eurochlor [19] provides insights into the performance change over time in chlor-alkali plants in Europe. The replacement of mercury cells with membrane cells, reduction in energy consumption, the extent of by-product hydrogen utilisation, trends in mercury emissions, and reduction of chlorine transportation by road and rail are reported over the period 2001–17.

7.9 Factors influencing learning

Whilst S curves are valuable in demonstrating quantitatively the rates of progress in performance embodied in process plants, it is important to reflect on the wider influences on learning, including those which are difficult to quantify or model.

In earlier chapters, in the contexts of market evaluation, capital and operating cost estimation, and profitability assessment, the importance of learning from a wide and diverse spectrum of experienced personnel, as well as from previous projects, has been emphasised.

In these contexts, the importance of environmental, safety, and sustainability assessments as part of the wider input into project design and evaluation have also been emphasised. Whilst it is difficult to quantify progress in these areas mathematically, it is possible to identify progress over time. Tables 7.5–7.7 demonstrate examples of some important timelines in environmental, safety, and sustainability awareness. Just as this knowledge takes time to emerge and gain acceptance by legislators, government policymakers, and industrial practitioners, so also the practice of translating this knowledge effectively into design, operation, and technology evaluation takes time. There is thus a path of time-dependent improvement extending through the industry, government (in legislation and policy development), and the wider community.

7.9.1 Environmental learning

Table 7.5 lists just some of the examples of past events leading to learning about environmental impacts. Most realisations of environmental damage emerge some years after damage has occurred; the related time period varies with the particular impact. Categories of environmental damage impacts are diverse including the well-known examples of global warming, acidification, eutrophication, ozone layer depletion, photochemical smog, and resource depletion, but there is a wide spectrum of toxicity related impacts affecting humans, animals, birds, marine life, and plants.

Table 7.5 Some important environmental timelines.

Year	Event	Details surrounding event	Some outcomes
1952	Great London Smog	Approximately 3000 people died arising from excessive concentrations of smoke and acid gas in atmosphere, derived from domestic coal combustion	Clean air legislation in United Kingdom (1956, 1968) making use of smokeless fuel obligatory
1962	Publication of 'Silent Spring' by Rachel Carson	Book focussing on effects of certain chemicals (particularly pesticides) in the environment	DDT insecticide banned from agricultural use in United States
1972	Publication of Club of Rome report 'The Limits to Growth'	Report focussing on effects of growth in population and world economy on waste emissions and resource consumption	The report (Meadows et al.) stimulated worldwide attention and concern on sustainability outcomes
1976	IChemE working party report on material and energy resources	Report focussed on potential resource shortages and their economic, social, and political implications	Issues included energy source options, materials recycling, substitution options, esp. in United Kingdom
1987	Montreal Protocol on substances leading to ozone layer depletion	In all, 23 nations agreed to cut usage of key chlorofluorocarbons (CFCs) by 50% by 1999	1990 London Amendment of 93 nations agreed to cease production of CFCs and most halons by 2000
1989	Exxon Valdez Crude oil spill	Ship ran aground at Alaska rupturing eight cargo tanks and releasing 258,000 barrels of oil. Extensive coastline damage and loss of marine life	US Oil Pollution Act (1990) increased penalties for ship oil spills and required double-hull tanker design
1992	Basel Convention on hazardous waste transport	Response to reports of toxic waste exports to developing world countries for disposal	Control of transboundary movements of hazardous wastes and their disposal
1992	Centre of Environmental Science, Leiden	Publications including 'Environmental Life Cycle Assessment of Products'	Further publications, including Guinee 'Handbook of Life Cycle Assessment 2002'
1997	Kyoto Protocol	Drafting of legally binding document on greenhouse gas abatement	Developments in details and global commitments leading to Paris climate change agreement

Continued

Table 7.5 Some important environmental timelines—*cont'd*

Year	Event	Details surrounding event	Some outcomes
2003	Sandestin, United States	First Conference on Green Engineering	Agreement on nine principles on green engineering
2009	Copenhagen	United Nations Climate Change Conference	UK introduces price of £16 per tonne on CO_2 emissions in 2013
2015	Paris Climate Change Agreement	UNFCC (United Nations Framework on Climate Change). Agreement by parties to intensify actions and investments towards a sustainable low carbon future	Commitments to limit global temperature rise and report emissions and emissions reduction strategies

In the process industries, learning about environmental impacts has led to technology improvements in processes and equipment, but also to some abrupt changes leading to the cessation of products in the market place and closure of manufacturing plants. Some examples of radical cessation include the elimination of CFC and halons production in the wake of the Montreal protocol. Other marked changes have been more gradual, for example:

- progressive closure of mercury cell chlor-alkali plants and replacement with membrane cell plants
- adoption of the treatment of sulphur dioxide emissions from metallurgical smelters
- transition from coal to gas for industrial and domestic steam and electricity generation

Over the period 2001–17 the percentage of mercury cell plants in chlor-alkali production in Europe is reported to have decreased from 60% to near zero whilst the percentage of membrane cell plants has increased from 20% to 80% [19]. Mercury emissions from production sites, which previously used mercury cells are reported to have been reduced from 2.5 g Hg per tonne Cl_2 capacity in 1995 to 0.68 g Hg per tonne Cl_2 capacity in 2017.

Mercury is a pollutant arising from a wide diversity of industrial sources, but as well as arising from mercury cell use in chlor-alkali production, it appears as a contaminant of bauxite in alumina refining, in natural gas and crude oil in hydrocarbon refining, in the treatment of mineral sulphide ores, in ash from some coals, and so on.

The economic consequences of pollutant emissions from process plants include site remediation but also increased costs to industrial manufacturers through the need to improve processes and in some cases replace plants. Some attempts have been made to put economic costs on more common emissions for example \$/t CO_2 or \$/t SO_2 emitted, but whilst potentially useful in weighing up mitigation options, these values are imprecise as the severity of impacts varies globally.

There are regular cases of emerging concern about the environmental and human safety aspects of established products. Recent examples include safety concerns about the use of the herbicide glyphosate, used in agriculture to control a wide variety of leafy weeds.

Such examples highlight both the relevance of time-dependent learning and a degree of related uncertainty and risk.

7.9.2 Process safety learning

Important contributions have been made by scientists and engineers in various capacities to improve our knowledge of process safety. Unfortunately, much of our knowledge has stemmed from the formal enquiries derived from major accidents, which have had serious social and economic consequences. Table 7.6 lists some serious past events and their contribution to learning about process safety. Safety initiatives in the design of new plants may result in additional capital expenditure through improved process control provisions and through additional design time spent in safety studies, such as hazard identification and HAZOP. Poor safety records of operating companies can lead to increased insurance premiums. Demolition and rebuilding of plants resulting from major accidents incur massive capital expenditure. Payouts are required for personnel killed or hurt. Income losses incurred directly from interruptions in production and indirectly from a decline in business reputation are significant.

Reports on the Buncefield incident estimated £900 million in compensation claims for site operators and aviation expenses, as well as £70 million for site rebuilding costs [20].

The Longford gas explosion in Victoria in 1998 resulted in the loss of natural gas supply to industrial, commercial, and domestic users for some 2 weeks. Large economic business losses were reported by gas users in industrial and commercial companies. Class action initiated in 2002 resulted in payouts of \$32.5 million for those suffering property damage [21].

Reviews of major accidents have identified frequently occurring shortcomings:

- poor design features of plants and facilities
- excessive inventories of hazardous materials
- inadequate training of process operators and/or operational staff
- inadequate emergency response procedures and preparation

Table 7.6 Some timelines in development of some process safety principles.

Year	Event	Details surrounding event	Some learning outcomes
1974	Flixborough, United Kingdom Explosion at caprolactum plant	Release of 30 tonnes of cyclohexane liquid caused a vapour cloud explosion. On-site, the entire plant and buildings were destroyed, 28 people died and 36 were injured; off-site, buildings and civilians were damaged	Recognition of importance of critical reviews of plant modifications during plant operating life, and of hazardous material inventories
1976	Seveso, Italy release of toxic TCDD from hexachlorophene plant	Loss of control on a reactor led to a bursting disc rupture with release of TCDD; rain and TCCD dispersion caused contamination of the surrounding district, with resulting health issues	Recognition of importance of containment of releases on plant, as well as effective communication and evacuation in event of an emergency
1984	Bhopal India MIC (methyl isocyanate) release from pesticide plant	Release of some 25 tonnes of MIC led to death of some 2000 people and injuries to some 20,000 living nearby	Principles of inherent safety, inventory management, and safe buffer zones emphasised in accident review
1986	Chernobyl, Ukraine	Operational trial at nuclear power plant failed and ended in release of radioactive material	Recognition of importance of prompt implementation of emergency measures
1986	Basel Switzerland Warehouse fire	Fire caused offensive smoke with local population advised to stay indoors. Toxic firefighting chemicals washed into Rhine river causing death to local fish	Recognition of importance of warehouse regulation, including limits on certain chemicals. Importance of providing basins for catchwater
1988	Piper Alpha North Sea	Gas explosion on offshore oil and gas platform. 167 workers were killed	Recommendations on permit to work procedures, provisions of fire protection on platforms, and emergency escape routes

Table 7.6 Some timelines in development of some process safety principles—*cont'd*

Year	Event	Details surrounding event	Some learning outcomes
1998	Longford, Victoria Australia	Onshore gas explosion killing two workers and injuring eight. Gas supply to Victoria was halted for 2 weeks	Royal commission of enquiry with major hazard facility legislation introduced
2005	Buncefield incident, Hertfordshire, United Kingdom	Initial explosion and consequent fire engulfing oil storage depot. In all, 40 people were injured with significant damage to offsite buildings	Recommendations on improved design and operation of fuel storage sites

7.9.3 Sustainability learning

The concept of sustainability had its origins in concerns over various forms of environmental damage and their long-term consequences. The concept has been widened to include social aspects including consideration and engagement of stakeholders. Safety is a dominant consideration of social aspects. Sustainability also relates to economics: this is both in the traditional approach to economic evaluation with emphasis on robustness to external influences over the project life and also on external costs arising from failure in environmental, social or safety performance. A detailed account of the origins of sustainability is provided by Perdan [22].

Table 7.7 summarises some timelines in establishing and adopting sustainability principles. Whilst the Brundtland report was a significant milestone, there were earlier developments, which paved the way. For example, foundations for the Earth Summit at Rio de Janeiro in 1992 can be traced back to the UN Conference on Human Environment at Stockholm in 1972 at which the United Nations Environment Program (UNEP) was established. In a briefing paper entitled 'Towards Earth Summit in 2002' [23], eight major steps prior to that event were identified, emphasising that behind the well-known timelines, there is a history of time-dependent learning and development. A detailed account of the origins of sustainability is provided by Perdan [22]. Implications for process engineering are explored by Brennan [24].

Table 7.7 Some sustainability timelines.

Year	Event or source	Details surrounding event	Some outcomes
1987	Brundtland Report	A total of 21 country members of World Commission on Environment and Development agreed on sustainability issues	Emphasis on ecologically sustainable development and preserving biodiversity
1992	Earth Summit Rio de Janeiro	Agreement reached to a global action plan, Agenda 21, with 27 supporting principles	Some better known principles included • Precautionary principle • Polluter pays principle
2000	Global Reporting Initiative (GRI) guidelines first published	GRI network established, with guidelines progressively refined sine 2000	Annual sustainability reporting widely practised by companies and organisations
2002	Johannesburg Summit	Tens of thousands of participants from a diversity of government and business organisations. A wide range of issues were discussed	Agreement on targets regarding poverty eradication, water and sanitation, energy supply, chemicals use, and natural resource management
2002	IChemE Sustainability metrics published	Publication of a set of indicators for measuring the sustainability of an operating unit	A range of metrics are provided covering environmental, economic, and social indicators
2015	IChemE	First issue of Journal of Sustainable Production and Consumption	A wide range of papers on energy, waste management, renewable resources, etc., in different industry and global contexts

Sustainability has been incorporated into industrial practice, professional societies and education, including those related to chemical engineering. Signs of this include

- regular publishing of sustainability reports by companies in conjunction with their annual reports to shareholders
- publishing of chemical engineering textbooks focussing on sustainability
- publication of journals related to sustainability
- promotion of sustainability within professional society reports, publications, and conferences

A significant link between sustainability and economics in the process industries includes the extended framework for evaluation extending to

- consideration of economic, safety, environmental, and social criteria as essential and interdependent inputs in project design and evaluation
- widening stakeholder involvement in the consultation process
- adopting extended system boundaries and time horizons in evaluating impacts and benefits from both existing operations and planned projects

Table 7.8 summarises some key interactions between economic evaluation and safety, environmental, and sustainability criteria.

Table 7.8 Interconnections between economic and other criteria.

Criterion	Economic implications
Safety	Impacts of failures • Compensation costs to injured personnel or damaged property • Direct (from plant shutdown) or indirect (from bad publicity) loss of business • Increase in subsequent costs of plant insurance
	Costs arising from • Increased number of qualified personnel and increased time required for evaluation at early (inherent safety) and later (HAZOP) stages of project development • Potential increase in capital cost of plant due to improved instrumentation and control, improved plant layout, or improved equipment design, including materials of construction
Environmental	Impacts of failures • Loss of production and related sales revenue resulting from shutdown for plant modification to meet required standards • Fines due to infringement of emission regulations • External costs arising from damage to the environment
	Costs arising from • Increased capital and operating costs required to meet improved standards in emissions reduction and resource consumption, e.g. through heat integration, water integration • Increased number of qualified personnel and increased time required for evaluation during project development • Imposed taxes (e.g. carbon tax on greenhouse gas emissions)
Sustainability	Need for robustness to change and related adaptability in relation to change in • Government regulation • Market factors, including for example need for improved product quality or packaging requirements, reduction in product demand resulting from recycling • Technology

7.9.4 Ongoing challenges in sustainability

Many important environmental and sustainability challenges remain in 2020 and beyond, but include

- strategies to reduce and adapt to global warming and related climate change
- the need for more effective recycling of used metals, polymers, glass, paper, and other materials
- the need to address specific pollution issues such as microplastics in the marine environment

These challenges are complex and require change on national and global scales, impacting not only on the environment if unresolved, but demanding change in social behaviour, government policy, regulation, and industrial practice.

Global warming was just one of the environmental impact categories documented in the first edition of the CML handbook on life cycle assessment published in 1992 [25]. The first published papers in the chemical engineering literature incorporating global warming as part of environmental assessment occurred in the mid-1990s. Despite reports by the International Panel for Climate Change (IPCC) highlighting the implications of global warming for rising sea levels and intensification of water cycles, there was considerable scepticism amongst the wider population and many political leaders in the 1990s. This scepticism has gradually been reduced particularly in the light of increased evidence more recently of increased frequency of heatwaves, bushfires, storms, and floods as well as coastal erosion, all with a spectrum of major social and economic impacts.

Global greenhouse gas emissions stem from electricity generation, transport, manufacturing industry, and other sources. Regular reports by the International Energy Association provide data and trends in sources of greenhouse gas emissions, as well as trends in relative contributions. A comprehensive report on potential emissions reduction in the industry [26] identified iron and steel, cement, chemicals (including plastics, fertilisers, and others), pulp and paper, nonferrous (aluminium and others) and food processing, textiles and leather, and mining as major contributors. The report identified six main categories of mitigation options but commented on a lack of comparable, comprehensive, detailed quantitative information, and literature on costs associated with each of the mitigation options. This highlights related economic challenges, including those related specifically to process industries.

The International Monetary Fund (IMF) has recommended carbon taxes as the most efficient means of reducing greenhouse gas emissions. The Organisation for Economic Cooperation and Development (OECD) provides regular reports on approaches to carbon taxing by different countries.

7.10 Introduction of technological change into new and operating plants

Despite the availability of improved technology, it is not always judged appropriate to introduce technological change. In the case of new plants, the risk of unproven technology or the desire to maintain a particular business liaison with a technology licensor may be contributing factors. In some cases, there may be reservations about the adequacy in skills of operating and maintenance personnel to cope with the technology. For example, in 1963 when 130 kA cells for aluminium smelters were available, cells in the 70–100 kA range were believed most economical because of operating difficulties at higher currents; cells of 50–80 kA were recommended as being best suited for developing countries at that time because of greater ease in operation [27].

In the case of introducing technological change into operating plants, the constraints of the existing plant and its process technology may be very considerable. Technological change may be so radical that it is technically feasible only when incorporated in an entirely new plant. In some cases, the introduction of new technology alongside older technology leads to problems of plant operability. In other cases, the incorporation of technological change may involve such extensive revamping of plant that the required shutdown time for the plant presents excessive difficulties and costs. In yet another set of cases, the incorporation of technological change is technically feasible and desirable but cannot be justified using conventional economic criteria.

When considering the incorporation of improved technology into an operating plant, the perceived ultimate capacity and life of the plant influence the estimated profitability. Frequently there is no clear or agreed view on the ultimate capacity and life of a plant, especially for plants which can be expanded by parallel addition. Case studies of Australian plants have indicated a consistent trend towards underestimating the ultimate life and capacity of plants [28].

Introduction of technological change into an operating plant often brings benefits of additional plant capacity as well as a reduction in production cost. Thus the introduction of metal anodes into operating chlorine plants to replace graphite anodes offered cost savings in energy consumption, labour, and chemical reagents, and also provided an increase in plant capacity. Where market growth rewards such capacity growth with increased revenue, the economic viability is clearly enhanced.

There are many examples of a successful introduction of technological change into operating plants. One common example is the introduction of improved instrumentation and control into older plants with benefits of improved product quality, and savings in labour, raw materials, and utilities. There are many other examples in plants, which have increased capacity during operating life [29]; in such plants, technological change is frequently incorporated directly into plant expansions.

7.11 Technological forecasting

A number of approaches to technological forecasting have been adopted. De la Mare in his book 'Manufacturing Systems Economics' [3] lists and briefly describes six techniques:

- expert opinion analysis or the Delphi technique
- time-trend analysis
- predicting by analogy (to another technology)

- morphological analysis
- conditional demand analysis
- scenario writing

Any rigorous approach to forecasting demands a clear understanding of the present position of the technology, tracing previous progress for the technology, and assessing the potential for further progress. For a process technology a number of performance indicators are quantifiable, for example:

- raw material consumption/t product
- energy consumption/t product
- persons employed/t product
- fixed capital investment/t product

For technologies involving production of a commodity product of well-defined composition these parameters can be amalgamated into a single production cost function (or its inverse, a production cost productivity function). This approach has been attempted for aluminium smelting, chlorine/caustic soda manufacture, ethylene production, and catalytic cracking of hydrocarbons [16].

A number of performance indicators for process technologies are less readily quantifiable. These include environmental, health and safety standards, the ability to process a range of feedstock compositions, flexibility in relation to co-product production, and (where applicable) variations in product quality.

By adopting a dominant performance parameter for the technology, or a performance rating based on weightings of several performance indicators, progress can be traced over the history of the technology. Assuming the life cycle curve follows the commonly observed sigmoid curve, an ultimate limit to technology development can be postulated, which assists in estimating the degree of maturity of the technology (or its 'position' on the 'S curve'). Limits may be set by thermodynamic or stoichiometric considerations, but also by less readily assessable practical or economic considerations.

Fig. 7.5 shows a plot of production cost productivity versus time for aluminium smelting technology based on the Hall–Heroult process over the previous century. Progress has been traced in raw materials consumption, energy consumption, number of employees, and fixed capital, all per tonne of aluminium. These parameters have been amalgamated to provide a production cost function using Eq. (5.1) and estimates of typical unit costs. The production cost function has then been plotted in its inverse form as a productivity parameter. Stoichiometric and thermodynamic limits have been assigned for raw materials (alumina and carbon) and energy consumption, and less rigorous estimates postulated for personnel and fixed capital. These limits allow an estimate of the ultimate limit to production cost productivity. This, in turn, allows the data set to be fitted to sigmoid curve models; the aluminium smelting data, in this case, showed a close fit to the logistic model.

In assessing the potential for further progress, one needs to allow for the potential for new advances derived from technologies outside the immediate process technology concerned. Many of the advances in process technologies

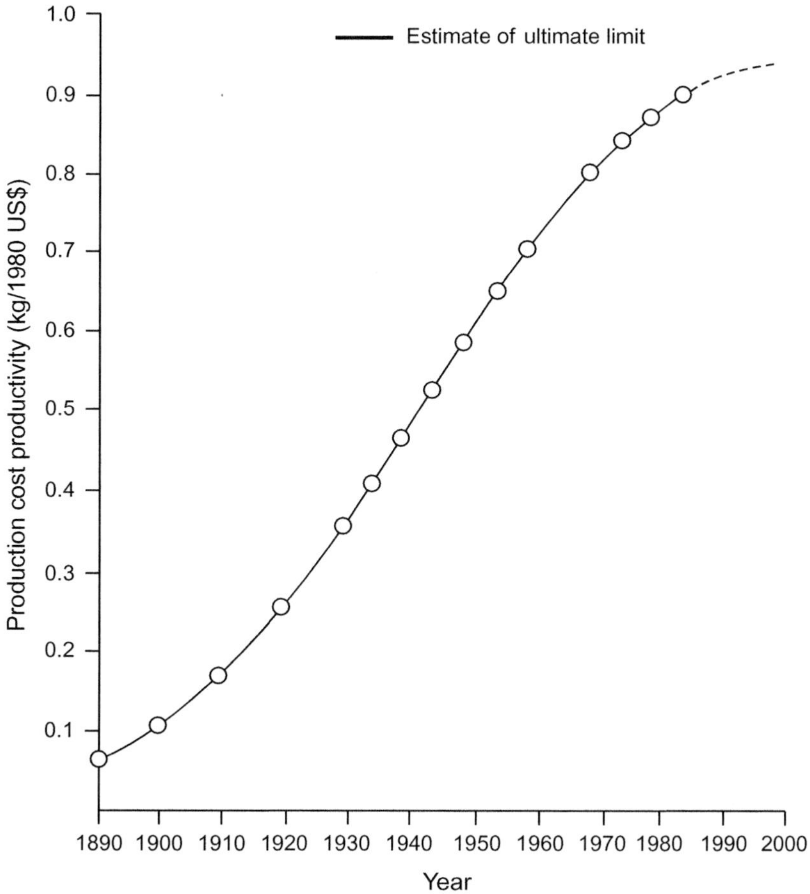

FIG. 7.5

Productivity development for aluminium smelting [16].

have been derived from advances in materials of construction for instance. It is likely that much of the current research in biotechnology, microelectronics, and new materials will provide benefits, which can be applied in process technology development. In the case of aluminium smelting, for example emerging technologies include the development of inert, carbon-free anodes, and new cathode materials.

7.12 Use of S curves

Apart from providing a valuable understanding of the evolution, development, and potential for process technologies, S curves have some important practical uses in business planning and management.

Firstly, S curves are useful in assessing a plant's performance relative to other plants including new plants using the best available technology and scale. Given the age of a plant, the S curve will provide a good indication of the performance of the plant when designed and constructed. With judgement, an allowance can be made for any subsequent expansion and/or technological improvement of the plant, aided by knowledge of its operating history. The adjusted performance of the plant can then be compared with that of competitors' plants using the same basic process route, taking into consideration their age, capacity, and operating history.

Another use of S curves is in R&D management. If a particular technology is reaching maturity, continued research may provide only minor gains for much research effort and cost. A better strategy is to put effort into developing an alternative technology, which whilst inferior at the time may be at a much earlier phase of its S curve and have the potential to surpass the established technology. A useful insight into factors influencing innovation success into new chemical products for different applications is provided by Miremadi et al. [30] where the time required for commercialisation as well as profitability considerations are discussed.

Quite apart from quantitative applications, the basic concept of S curves is of fundamental importance highlighting the role of time and limits in achievement. This applies to learning on technology, safety, environmental and sustainability issues as well as their management and regulation, and the implications for ongoing capital investment and operating costs.

References

[1] R.U. Ayres, Technological Forecasting and Long-Range Planning, McGraw-Hill, New York, 1969.

[2] D.R. Coates, Technological forecasting and the planning of R and D, in: R.V. Arnfield (Ed.), Technological Forecasting, Edinburgh University Press, Edinburgh, UK, 1969, pp. 83–96.

[3] R.F. de la Mare, Manufacturing Systems Economics, Holt, Rinehard and Winston, London, UK, 1982.

[4] D.H. Allen, The chemical industry and technological innovation—an economic perspective, in: Proceedings of Chemeca 86 Australasian Chemical Chemical Engineering Conference, 1986, pp. 28–32.

[5] W.H.C. Simmonds, Analysis of industrial behaviour and its use in technological forecasting, in: Bright, Schoeman (Eds.), A Guide to Practical Technological Forecasting, Prentice-Hall, New York, 1973.

[6] D.J. Brennan, An alternative approach to modelling performance improvement in the process industries, Chem. Eng. Aust. CE12 (4) (1987) 7–12.

[7] D.F. Ball, A.W. Pearson, The future size of process plant, Long Range Plan. (1976) 18–28.

[8] C.F. Pratten, Economies of Scale in Manufacturing Industry, Cambridge University Press, London, 1971.

[9] C. Ramshaw, Higee distillation—an example of process intensification, Chem. Eng. (389) (1983) 13–14.
[10] Euro Chlor, Chlor-Alkali Industry Review, https:/www.eurochlor.org/wp-content/uploads/2019/05/euro_chlor_industry_review_FINAL.pdf, 2017/2018.
[11] D.J. Brennan, Evaluating scale economies in process plants: a review, Trans. IChemE 70 (Part A) (1992) 516–526.
[12] R. Malpas, Chemical technology—scaling greater heights in the next ten years? Chem. Ind. (5th February) (1977) 11–118.
[13] W.B. Hirschmann, The learning curve, Chem. Eng. 71 (7) (1964) 95–100.
[14] Boston Consulting Group, Perspectives on Experience, Boston Consulting Group, MA, 1970.
[15] J.H. Taylor, P.J. Craven, Experience curves for chemicals, Process Econ. Int. 1 (1) (1978) 13–19, 28.
[16] D.J. Brennan, Tracing S-curves for process technologies, Process Econ. Int. 7 (2) (1988) 66–71.
[17] M.P. Decleva, Technological Change in Oxygen Production (M. Eng. Sc thesis), Monash University, 1992.
[18] J.L. Enos, Petroleum Progress and Profits—A History of Process Innovation, MIT Press, Cambridge, MA, 1962.
[19] Eurochlor, Chlor-Alkali Industry Review, www.chlorineindustryreview.com, 2017/2018.
[20] Buncefield Major Incident Investigation Board, The Buncefield Incident 11 December 2005, in: The final report of the Major Incident Investigation Board, vol. 1, Health and Safety Executive, Bootle, UK, 2008.
[21] Longford Gas Plant Class Action, https:/www.mauriceblackburn.com.au/class-actions/past-class-actions/longford-gas-plant-class-action/.
[22] S. Perdan, Introduction to sustainable development, in: A. Azapagic, S. Perdan, R. Clift (Eds.), Sustainable Development in Practice, Wiley, Chichester, UK, 2004.
[23] R. Gardiner, Towards Earth Summit 2002. Briefing Paper, www.earthsummit2002.org, 2002.
[24] D. Brennan, Sustainable Process Engineering, Pan Stanford, Singapore, 2012.
[25] J.B. Guinee, Handbook on Life Cycle Assessment, Centre of Environmental Science/Kluwer Academic Publishers, Leiden/Dordrecht, 2002.
[26] M. Fischedick, et al., Industry (Chapter 10), in: Climate Change 2014: Mitigation of Climate Change. Contribution of Working Group III to the Fifth Assessment Report of the Intergovernmental Panel on Climate Change, 2014.
[27] Department of Economic and Social Affairs, Pre-Investment Data for the Aluminium Industry, United Nations, New York, 1966.
[28] D.J. Brennan, Capacity and Performance Profiles of Process Plants, (PhD thesis),University of Melbourne, Melbourne, Australia, 1986.
[29] D.J. Brennan, G.K. Stephens, The extent and rate of capacity increase in process plants, Process Econ. Int. V (3 and 4) (1985) 15–25.
[30] M. Miremadi, C. Musso, J. Oxgaard, Chemical innovation: an investment for the ages, McKinsey on Chemicals (May) (2013).

Further reading

IEA Energy Technology Systems Analysis Programme Technology Brief, Aluminium Production, March 2012.

CHAPTER

Capital investment decisions

8

A reasonable estimate of economic organisation must allow for the fact that, unless industry is to be paralysed by recurrent revolts on the part of outraged human nature, it must satisfy criteria which are not purely economic.

R.H. Tawney (1880–1962), British economic historian in 'Religion and the Rise of Capitalism' (1926)

Chapter outline

Process Industry Economics. https://doi.org/10.1016/B978-0-12-819466-9.00008-0

8.1 Expenditure proposals

Before an operating company can have a project implemented, approval must be granted from within the company. This approval is normally given by the company's board of directors and represents not only the decision to spend the required capital, but also to accept responsibility for the project as a whole over its entire life, with all its financial, technological, safety, environmental, social, and political implications. Thus the commitment to proceed with projects is properly the responsibility of the highest level of management, consistent with the company's responsibility to its employees and shareholders and to the wider community. Often this responsibility is delegated for smaller expenditure projects to senior levels of management, with an approval hierarchy, which corresponds to the magnitude of funds for the project.

It should be emphasised, however, that small projects still require careful assessment of their technical, economic, safety, and environmental merits.

The proposal to spend capital for a process industry project must be integrated with a company's overall capital expenditure schedule, planned well ahead of project sanction, and integrated within the company's broad financial plan.

An expenditure proposal will normally have multiple parts:

- an estimate of the capital requirements (fixed, working, and land) for the project
- an estimate of the operating costs for the project
- details of the project including the site, feedstock, product and by-products, its integration with other processing facilities, storages, transport, plant production capacity, utility requirements, investment period, and operating life
- supporting technical documentation including process description, process flowsheets, equipment specifications, preliminary engineering flowsheets, and preliminary layout drawings
- an evaluation of the market for products
- details of the proposed means of financing the project
- an evaluation of the profitability of the project, usually based on cash flow projections, including the inherent uncertainties and financial risks
- a technical safety assessment for the project
- a technical environmental assessment for the project

The required accuracy for the fixed capital estimate for sanction purposes is normally $\pm 10\%$ (see Section 3.3.1 for further discussion). This accuracy demands a corresponding degree of project definition and engineering design. There will always be elements of uncertainty affecting the project. An unduly optimistic and noncritical approach will have little chance of success; at the same time, an excessively pessimistic and cautious approach is unlikely to see the project approved.

The quality of the proposal documentation as a communication between those developing and engineering the project and the approving authority is of key importance. The board of directors normally meets at scheduled times and proposals must be complete and prepared on time. Most companies will document expenditure proposal guidelines. There should be ample opportunity for the necessary preauthorisation communication between the board and the project team. Communication needs to be at the appropriate level of detail, recognising that board members are from a range of commercial or technical disciplines, and have less familiarity with project details than the project team members.

The expenditure proposal is very much a financial undertaking and must satisfy the scrutiny of the company's accountants. In the later stages of project development prior to sanction, consultation of technical personnel with accounting personnel is essential.

In nearly all cases, even where technical personnel are experienced in preparing proposals, accounting advice will be necessary. In any event, the expenditure proposal needs the consent of the company accountant (or delegate) before being submitted to the board. Consent of the company's operations manager will likewise normally be a requirement before submission. An operations representative would normally be a member of the project team.

Besides convincing all levels of responsibility within a company, it is important to convince the external community of a project's value. Projects impact on communities in a variety of ways—financially, employment wise, and environmentally for example and communities have perceptions about those impacts. These perceptions may be soundly or poorly grounded. There is a responsibility for companies to explain the essence and merits of projects to communities, especially those communities most directly affected. Neglecting these responsibilities may jeopardise a worthwhile project's chance of implementation. For major projects, the cost of this activity can be significant.

8.2 Shaping and scoping projects

In the cost estimation and economic evaluation of projects, it is essential to clearly define the boundary limits for the investment, operating costs, and product selling prices (see discussion in Sections 3.3 and 4.1). The timing for project approval, design, construction, commissioning, and subsequent operation must be defined. Production capacity must be stipulated. Utility requirements, by-products and effluents must be defined. The interaction of the project with other plants and industrial systems must be examined. Revamp projects require special attention in this regard (see Section 3.9).

Definition of boundaries is important in the context of inside and outside battery limits investments and also to raw materials supply and product despatch where the assumed prices must match the point and mode of supply. For example, the estimated selling price of a product to customers may correspond to supply from a warehouse;

in this case, the transportation cost from the manufacturing plant to the warehouse, the fixed capital cost of the warehouse facility, and the contribution to working capital of the warehouse stocks must be accounted for. Such system boundaries and any associated gaps in definition need to be identified early in a project. Similar considerations apply to the supply of raw material from the ship to dock storage and subsequent road transport to storage at the manufacturing site.

Another important issue lies in accounting for effluents from a plant. For example, a processing unit within a petroleum refinery such as a hydrotreater or an alkylation unit may generate several effluents which are best treated in different parts of the refinery outside the processing unit's battery limits. Effluents might include a gas stream containing combustibles for flaring, a gas stream containing traces of hydrogen sulphide for incineration, a sour water stream, and a loaded amine stream for regeneration. In estimating the operating costs for such a processing unit, effluents must be clearly defined in terms of mass flowrate and composition, their means of treatment, and the associated cost implications determined. A portion of the total treatment costs for these and similar effluents within the refinery must be apportioned to the processing unit. In some cases, spent catalysts may need to be removed from the refinery for regeneration or treatment; related costs must be accounted for.

Similar considerations apply to the supply of utilities such as steam, electricity, fuel, cooling water, and nitrogen. Where utilities are supplied by pipeline for example a clear definition of the pipeline boundary must be made so that the cost of any additional piping for the project may be borne by the project.

An important aspect of any project is a clear statement of the expected production capacity and performance of the plant. The annual production capacity of a plant can be regarded as the maximum daily capacity in tonnes per day of product multiplied by the equivalent availability of the plant in days per annum to produce at the maximum daily rate. The maximum daily capacity may be verified by short-term production trials lasting typically 1–3 days; these are usually carried out at start-up to verify both production capacity and performance. Performance trials conducted simultaneously with capacity trials are usually restricted to product quality and consumption of key raw materials and utilities. These are part of the contractual relationship between designers and operators of the plant. Equivalent availability allows for both off-line time of the plant and restrictions to achieving full plant capacity when the plant is on-line. Both the inherent operating characteristics of the plant concerned and its interaction with external systems must be considered.

Several different types of investment proposals are discussed in the following sections. In practice, there are projects where particular issues dominate and cases where multiple facets must be jointly considered.

8.3 Projects involving the replacement of plant and equipment

Replacement of facilities is considered when those facilities are judged to have reached the end of their economic life. Termination of the economic life of process plants may be brought about by a number of factors. The most telling reason for plant closure is the failure to generate cash; that is, the cash operating costs exceed the revenue which can be realised through product sales. In the case of an old plant, it may be difficult to reduce costs in a business environment where product selling prices are declining due to improved cost structures in newer plants.

Specific factors leading to termination of plant economic life include

- excessive maintenance costs or poor reliability resulting from age and wear of plant and equipment
- inability to meet environmental regulations
- deteriorating cost structure influenced for example by cost escalation of a specific feedstock
- inability to meet demands on product quality
- inability to show adequate return on investment on plant improvement projects
- inability to meet safety and health requirements

When plant closure occurs, replacement capacity for the product concerned may be taken up by the plant owner through new plant investment or expansion of an existing plant, either on the same site or a different site. Alternatively, replacement capacity may be taken up by a competitor.

Replacement of equipment within existing plants may be encouraged by

- excessive maintenance costs or poor reliability
- inadequate capacity
- the availability of superior technology in new equipment, offering improved operating cost, environmental or safety performance

Capital burdens, whilst significant, are not as potentially crippling as those for entire plant replacement and may bring major benefits. Adequate depreciation provisions must be made within operating budgets to ensure such replacements can be funded.

In justifying replacement decisions of equipment and plant, projections of costs must be made for the separate cases of replacement and nonreplacement. All cost and revenue implications including impacts on safety and the environment, on-line time, plant capacity, and product quality must be carefully examined. In equipment replacement within plants, an important but difficult estimate is the ultimate life of the plant. As the equipment is replaced, the potential life of the plant is prolonged. This may make the replacement of the entire plant more difficult in the longer term. The maintenance costs of old plants, particularly when required to meet demanding reliability and environmental targets, are likely to escalate markedly.

8.4 Environmental projects

Chapter 5 outlined the traditional approach to the economic evaluation of process industry projects within businesses based on cash flow analysis over the perceived time horizon for the project. In Chapter 7, the importance of learning has been reviewed, not only in relation to technology, but also in relation to environmental, safety, and sustainability impacts.

All process industry projects should be subject to a rigorous environmental assessment. With increased global attention being focussed on the quality of the natural environment, many process industry projects, as well as satisfying economic objectives, are also driven by a need to improve environmental performance. Such projects may be driven by

- effluents from plants discharged continuously or intermittently to air, water, or land, which have harmful effects
- risk of unacceptable effluent discharge arising from limited reliability or inadequate control of a process plant
- products whose use or ultimate disposal create an unacceptable waste
- commitment to reducing impacts on global warming
- concerns regarding the depletion of valuable resources consumed in processes

Environmentally driven projects may involve entire new plants or modifications to operating plants. The need to meet improved environmental standards may be driven by a combination of regulatory, community, or ethical pressures. Pressures may be derived from global concerns such as climate change related to global warming and/or to more regional concerns such as acidification from sulphur dioxide emissions. Failure to respond to these pressures carries risks of loss of market share, plant closures, loss of whole businesses, and damage to company image, all of which involve financial penalties, many of which are substantial. At the international level, relationships between nations and groups of nations may be seriously strained where some nations respond for example by imposing carbon taxes, whilst other nations take little or no action.

Environmental problems might be solved technically by a variety of approaches, for example:

- minimising the generation of waste at source by improved process design
- minimising consumption of raw materials and utilities
- making the process more energy efficient
- substitution of toxic catalysts, solvents, or other process materials by less harmful ones
- end of pipe solutions which treat effluent streams to produce a recyclable stream, a marketable product, or a harmless waste for disposal
- recycling products or their component materials to recover useful material or energy resources

All approaches involve technological and operational changes. These in turn frequently require incremental capital expenditure, involve changes in operating cost, and generate new environmental effects, which (in contrast with the fundamental environmental change being sought) must be minimal. In cases where not all the necessary technology is available, research and development programs and associated expenditures will be required affecting the timing and ultimate cost of the project.

8.4.1 Environmental assessment

The vast array of environmentally driven projects being tackled throughout the world's process industries requires a thorough economic and environmental assessment. Of the developed methods for environmental assessment, the most widely used involves various forms of life cycle assessment (LCA). LCA had its origins in evaluating the environmental merits of various consumer products, for example alternative forms of packaging, but has rapidly been extended to include all products relevant to the process industries. The method involves several distinct phases of analysis:

- definition of the *objective function*, for example different beverage containers of 250 mL capacity, or different insulating materials having a specific insulation duty, as distinct from necessarily 1 tonne of the material used for a product application.
- an '*inventory*' analysis which defines the inputs (materials and energy flows) and outputs (product and waste flows) at all phases of the product's life cycle beginning with the extraction of resources used for its manufacture, encompassing the various stages of processing, through to product assembly, use, and final disposal. Note that the term 'inventory' in this context is distinct from the concept of inventory commonly adopted to mean the quantity of a material stored in various ways in a process plant
- a *classificatio*n phase, which examines and quantifies the environmental implication of resource depletion or waste generation under approximately 14 categories of environmental impact including for example acidification, global warming, and resource depletion
- an *evaluation* phase, which attempts to amalgamate the various quantified impacts from the classification phase into a single index
- an *improvement analysis* phase, where the potential for reducing environmental impacts is examined in the light of the previous phases of analysis

Because of the large system boundary extending over the entire product life cycle (or from 'cradle to grave'), the analysis is very detailed and creates challenges in both obtaining the enormous volume of data necessary and making visible the basis and details of an LCA study. Nevertheless, many studies have been made, and many of these studies have been published (with varying degrees of visibility of the data and assumptions on which the study has been based).

The method is useful for analysing the environmental merits of manufactured chemical intermediates and alternative process routes, as subsets of the life cycle. It is well structured, has a sound scientific base, and is comprehensive, both in relation to its system boundary and the range of environmental effects examined. Many guidelines are published, for example those focussing on the methodology (a valuable guide is the handbook by CML [1]) and those focussing on applications in chemical engineering, for example Azapagic et al. [2] and Brennan [3].

Environmental assessments include numerical rankings to reflect the environmental impact, thus enabling comparison of alternative options, including the 'do-nothing' option. Other ranking approaches based on the weighting of relevant criteria have been used. Some of these criteria include community perception, which whilst possibly independent of scientific criteria, may strongly influence outcomes for project viability. Accounting for public perception can also be partly justified by the uncertainty often inherent in scientific measures of environmental impacts from specific mining or manufacturing projects.

8.4.2 Economic assessment of environmental effects or benefits

Costs of environmental damage from process industry projects, possibly neglected but often not foreseen at the investment phase, fall on the community at large. Because they fall outside of the accounting framework of the industrial company concerned, they are referred to as external costs or 'externalities'. Environmental externalities are potentially diverse, encompassing costs to human health, damage to buildings, degradation of water sources, damage to the animal, bird, and marine life and hence food sources, and so on. Estimates of externalities include those related to electricity generation emissions [4] and landfill disposal [5]. Estimates of externalities are inherently uncertain, partly since damage from impacts takes time to manifest itself, be understood and be evaluated, and partly since vulnerability to damage from impacts, even those which by nature are global, varies from one country and region to another. However, the estimates are important in guiding policy decisions at government and corporate levels on remedial actions. They can also influence related insurance costs.

In many cases, a corporate decision may be taken to invest in an environmental project because it is essential to the survival of the business. There will be benefits to the environment, company image, and survival of a product, plant, or business area. These benefits must be identified, but the assignment of precise financial values to these benefits may be difficult. In some cases, financial benefits may be expressed as expectations incorporating probability; in other cases as a reduction in financial risks. A useful reference book in this context is 'Financial Evaluation of Environmental Investments' by Moilanen and Martin [6].

The knowledge and skills required to estimate these financial benefits will be beyond the capability of most chemical engineers working in isolation from other professions. Nevertheless, the chemical engineer is required to consider options at the technical level, choose the best one, and justify its selection to management. Each option will have different levels of capital expenditure, operating costs, and secondary impacts. The least cost within the system boundary is one important criterion which the chemical engineer can evaluate as the minimum net present cost (before tax in the first instance, and after tax as options become better defined). The cost of capital for discounting should be selected in consultation with financial personnel and the estimated life of the project in consultation with management (see Sections 5.4 and 5.7.1 and 4.5 for further discussion). Since estimates of both the cost of capital and the project life are likely to be imprecise, the sensitivity of the net present cost to both variables should be examined (see Section 5.9). The definition of system boundary needs careful consideration and articulation. It may be a process plant, or section of a plant, but there will be streams entering and leaving the system which have wider environmental and economic implications.

8.4.3 Application of a cost-benefit approach in process industry cases

Cost-benefit analysis has been widely used in public expenditure projects to assess benefits to society in conjunction with costs. Cost-benefit analysis can also be applied in enviro-economic assessment whether in process development, selection, or design context. In these cases, economic parameters such as capital and operating costs for options can be weighed against the magnitude of environmental impacts determined using life cycle assessment. Some examples of this approach discussed by Brennan [3] include

- Process design options in sulphuric acid manufacture, where options include
  - increasing the number of reaction stages in converting SO_2 to SO_3
  - intermediate absorption of SO_3 within the reaction stages

These options provide benefits in overall conversion of feedstock to acid and reduced SO_2 emissions to the atmosphere, but increases in capital and operating costs, including those derived from energy consumption.

- Hydrotreating of diesel to reduce sulphur levels in diesel fuel product involving
  - provision of a hydrotreating unit with increased requirement for hydrogen supply and increased recovery of sulphur

This provides environmental benefits by reducing SO_2 emissions in a diversity of fuel combustion uses in transport and power generation but at the expense of increasing capital and operating costs in the refinery.

- Power generation from fossil fuels where
  - fuel costs are much lower for coal than natural gas, but the emission profile including CO_2, SO_2, and particulates is much higher for coal than gas
  - for the case of natural gas, lower gas consumption and lower CO_2 emissions per MWh are achieved for a combined cycle than for open cycle generation, but where combined cycle generation incurs higher capital costs

An additional complication for the investment decision arises from a changing horizon over time of progressively tighter emission limits. Is it better to meet existing emission limits, or should one meet projected tighter limits anticipated for the future whilst incurring higher costs? The incremental capital and operating costs in achieving better environmental performance levels become important and should be evaluated.

The case of sulphur levels in the diesel, which have become tighter in most countries over the last two to three decades, is just one example of a moving emissions target.

8.4.4 Some economic incentive instruments

A number of economic instruments are employed by regulatory authorities to improve environmental performance. These include

- **Emissions charges** on the discharge of pollutants into air, water, or soil and on noise generation. The charges are related to quantity and quality of pollutants and the related damage. An important example is the carbon tax.
- **User charges** which have a revenue-raising function and are related to costs involved in the collection, treatment, and disposal costs of waste and recovery of related administrative costs.
- **Product charges** levied on products which are environmentally harmful when used in production processes, consumed, or disposed of.
- **Tradeable permits** which enable the cost of pollution control to be borne by those polluters who can reduce pollution more economically. Following the allocation of permits, polluter A can sell permits to polluter B, who due to various circumstances (e.g. lack of access to key raw materials or facilities) finds pollutant treatment expensive.
- **Deposit refund systems** involving deposits paid on potentially polluting products. An example is the commonly adopted deposit system for returnable glass drink containers.

8.5 Safety projects

The potential loss of human life is an overriding consideration, which demands priority in projects. Loss or damage to human life and property have legal and financial consequences, but these are dwarfed by the human dimension in terms of the loss or damage incurred by those hurt directly and their close kin, and even the psychological damage suffered by those responsible.

Safety standards are built into projects and evaluated in terms of acceptable risks through hazard evaluation procedures. Additional capital may be required to achieve higher standards of safety or lower levels of risk and may be weighed against these perceived risk levels. The consequences of failure as well as their probability should be carefully evaluated.

Other benefits derived from changes made to achieve improved safety standards should be evaluated. These include reduced risk of environmental impact from a disaster or loss of containment, reduced fixed capital and working capital costs derived from reduced inventories, lower insurance premiums, improvements in process and plant reliability, and improvements in product quality, or plant performance resulting from improved process control.

Improved safety and environmental standards, including those derived from inherently safer and cleaner processes, are often important by-products of technological change embodied in new processes and plants. Opportunities to integrate improved health, safety, and environmental standards with technological change are thus an important priority in technology management. Since safety and environmental standards are so integral to technology development and plant design, it is often not feasible to extract the investment dedicated specifically to safety or to the environment within a project for incremental economic analysis. This is one reason why there is not a great deal published on the economics of safety or environmental standards in process plants. Safety and environmental standards are a precursor to project viability, not only from the project owner's and investor's viewpoints, but also from the viewpoints of lending authorities, governments and communities, and any party likely to be affected by the project over its total life.

8.6 Capacity planning

8.6.1 The need for capacity planning

The planning and sizing of manufacturing facilities for growing markets is an ongoing activity within process industries. It requires careful evaluation of the market, the technologies of existing and new plants, and the capital and operating costs of various options. Thus, it draws on much of the earlier chapters of this book. Capacity expansion for existing business areas is one of the major demands on capital funds and engineering skills within process industry enterprises.

Fig. 8.1 represents a situation where the demand for a product is approaching the existing manufacturing capacity of a company. Three basic strategic options are available to enable the manufacturing capacity to meet the demands for

- expansion of existing plant
- building of a new plant to provide additional capacity
- building of a new plant to provide replacement plus additional capacity

The choice of option is influenced by the availability of capital, by market factors, and by the competitiveness of existing operations. Some of the decisions involved in the options are illustrated in Fig. 8.2 as:

(a) expand old plant or build a new plant
(b) if expanding old plant, size of capacity increment and expansion strategy
(c) if building new plant, 'tight' design or design for future capacity growth

The first option of expanding an existing plant is encouraged by a market of low growth rate or small volume. Even with a high growth rate, a small market volume implies that the incremental growth in volume may be too small to support a new plant of sufficient capacity to be economically viable. A new plant sized to replace existing capacity plus capacity for projected market growth will be more competitive in technology and scale but makes large demands on capital; hence this strategy cannot be adopted too frequently by an operating company.

Expansion of an existing plant is attractive in terms of capital cost but tends to perpetuate the technology of the original plant. The safety and environmental

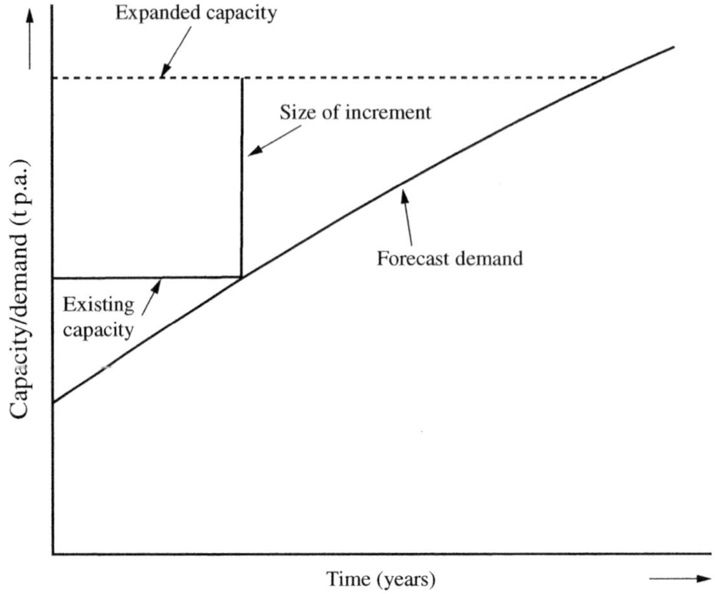

FIG. 8.1

Expansion problem—what size increment is needed to meet forecast demand?

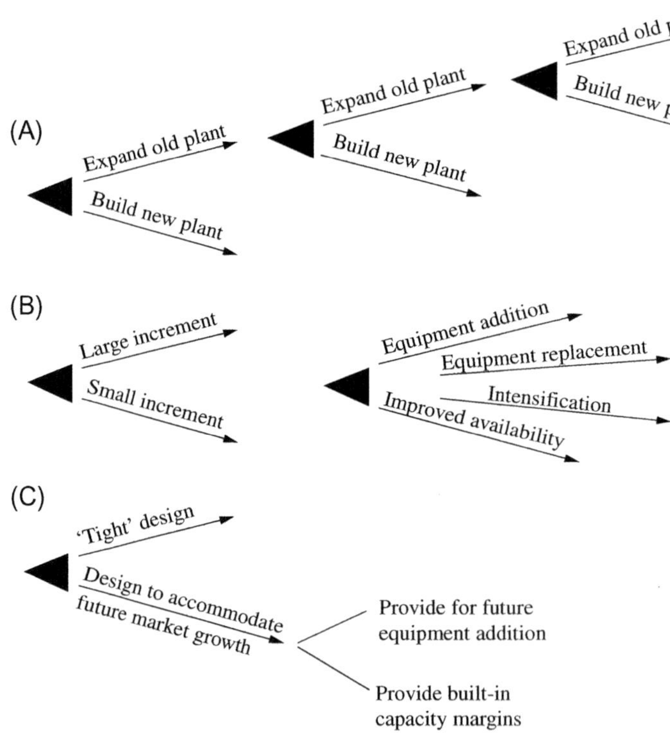

FIG. 8.2

Some strategic decisions in capacity planning [7]. (A) Expand old plant or build new plant; (B) if expanding old plant, size or capacity increment, and expansion strategy; (C) if building new plant, 'tight' design, or design for future capacity growth.

standards, personnel requirements, raw material and energy efficiencies, and product quality for the older plant all need to be monitored and compared with the performance of a new plant employing the best available technology and scale.

8.6.2 Technical and economic considerations in expanding an existing plant

Annual production capacity is the product of maximum daily capacity and equivalent availability, as discussed in Section 8.2. Equivalent availability can be increased by increasing storage facilities, improving the use of existing storage facilities, negotiating reduced frequency of vessel inspections by government authorities, or improving operating and maintenance procedures. Growth in annual plant capacity by this approach is likely to be limited in magnitude but may be achievable with minimal capital investment and may be of strategic importance.

Larger increases in production capacity are likely to require increase in the maximum daily capacity. These can be achieved by

- replacement of existing equipment with equipment of greater capacity
- addition of equipment, usually in parallel but possibly in series with existing equipment
- intensification of process or equipment

Intensification has been discussed in Section 7.5. Similar effects to intensification can be achieved, often with minimal technological change, by learning to exploit design margins during plant operation. Increased throughputs in process equipment can thus be achieved over time. Increased temperatures, pressure drops, or wear rates may result, and there may be penalties in raw materials or energy efficiencies, maintenance costs, and equipment life. These disadvantages, however, may be tolerable in the interests of increased production in the complete plant. This is largely an experimental and experience-dependent phenomenon.

The strategies of replacement, addition, and intensification can be adopted at various process steps within the plant as required, to increase overall plant capacity. A capacity distribution normally exists in a plant as illustrated in the simplified diagram in Fig. 8.3. Capacity distribution in a new plant results from attempts to accommodate design uncertainties and construction constraints. The extent of the margins tends to reflect the maturity of the process—uncertainties and hence margins are greater with new processes. The capacity distribution may change during plant life

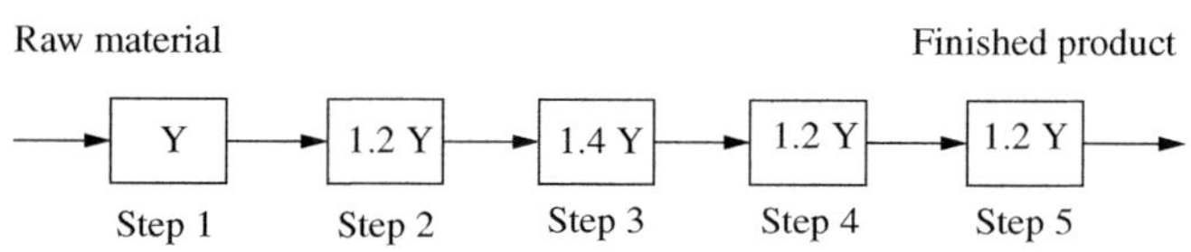

FIG. 8.3

Possible capacity distribution in a process plant [8].

resulting from changes in feedstock or product composition, technology, operating procedure or the plant itself. An important part of economic plant expansion is the identification and exploitation of such capacity margins at various steps within the process. For the distribution in Fig. 8.3 for example, a 20% increase in overall plant capacity is achievable by increasing capacity at Step 1.

Two basic types of expansion can be distinguished (see Fig. 8.4):

(a) **Parallel streamed expansion**, where parallel addition is made at each step within the process.

(b) **Constrained expansion** where parallel addition is precluded in at least one step within the processing sequence. In some cases, a single stream plant may retain its single stream structure throughout its life, despite having undergone several expansion steps.

Because of the freedom to add capacity when required, parallel streamed expansions can achieve a closer match between plant capacity and market demand than can constrained expansions. Parallel streamed expansions can also achieve higher rates of capacity growth (typically up to 10% per annum) than constrained expansions (typically limited to 4% per annum) [2].

Plants undergoing constrained expansions typically show a declining rate of capacity growth with time, consistent with progressively increased difficulty in exploiting capacity margins at successive expansions. For Australian plants undergoing constrained expansions in the 1960s and 1970s, Brennan and Stephens [8] found that Eq. (8.1) initially proposed by Malloy [9] provided a good fit for capacity growth phases.

$$(Q_t - Q_i)/(Q_u - Q_i) = 1 - \exp(-t/\tau) \tag{8.1}$$

where

Q_t = capacity at year t
Q_i = initial capacity
Q_u = ultimate capacity
τ = time constant (years)

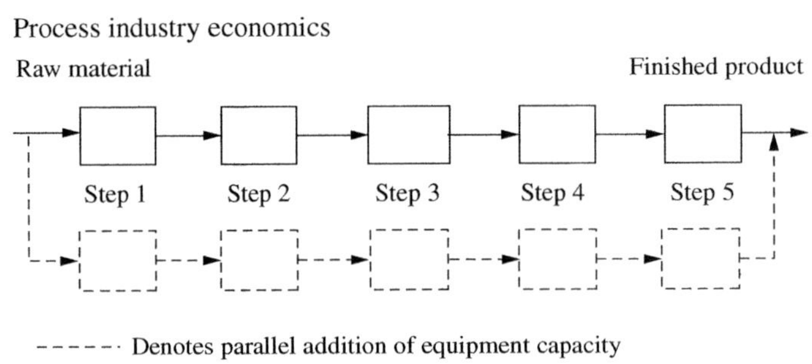

FIG. 8.4

Schematic outline of a parallel stream expansion.

Brennan and Stephens found mean values of $Q_u/Q_i = 1.6$ and $\tau = 2$ years, for capacity growth phases. Capacity growth frequently commences some years after start-up. In some plants, several capacity growth phases may occur; new growth phases are initiated by major technological change or major capital investment, which creates new capacity distributions.

Figs 8.5 and 8.6 show examples of capacity over time profiles for plants undergoing parallel streamed and constrained expansions respectively. Expansion of the aluminium smelter occurred through the parallel addition of potlines and transformer rectifiers, with the replacement of the initial anodes and (subsequently) of the initial potline. A potline comprises a large number of 'pots' or reactors where alumina is reduced to aluminium.

Expansion of the ethylene plant occurred partly through the parallel addition of cracking furnaces, but the downstream separation and purification plant remained essentially single stream. The cracking furnaces are high-temperature reactors in which naphtha is converted to ethylene and other products. Plant expansion occurred in three stages over a 5 year period. Some 16 additions of ancillary equipment items, 24 modifications to equipment, and process modifications enabling the capacity increase of selected equipment items were made. However, there were no parallel equipment additions to major cost items such as distillation columns, main gas, and refrigerant compressors. Further details of these and other plant expansions are provided in Ref. [8].

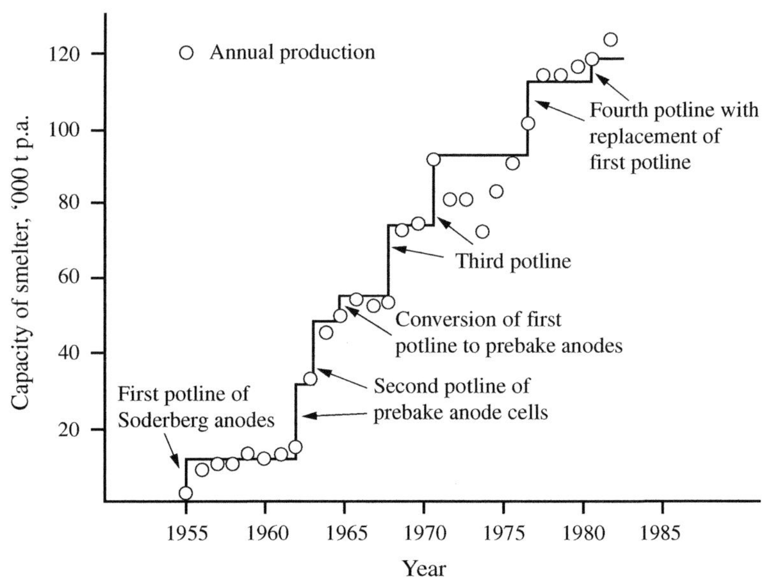

FIG. 8.5

Capacity history of an aluminium smelter [8].

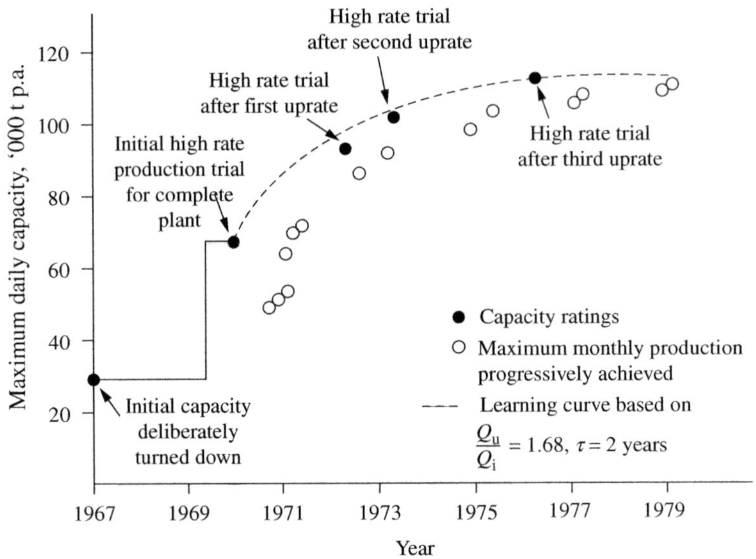

FIG. 8.6

Capacity history of an ethylene plant [8].

8.6.3 Cost and performance changes in expansions

Fixed capital costs for plants undergoing expansion can be correlated by the relationship

$$\sum I = k_i Q^b \tag{8.2}$$

where

$\sum I$ = cumulative fixed capital adjusted to a real basis spent on the plant
Q = plant production capacity
k_i = proportionality factor
b = exponent

The value of the exponent 'b' was found to average 0.64 for all plants examined in the studies undertaken by Brennan and Stephens [10]. Values for individual plants were similar to values indicative of static scale economies; for example a value of 0.75 was obtained for an aluminium smelter over six expansion steps, and 0.58 for a single stream vinyl chloride plant over four expansion steps.

Total numbers of personnel employed on expanding plants can likewise be correlated with capacity by the relationship

$$M = k_m Q^a \tag{8.3}$$

where

M = total number of persons employed
k_m = proportionality factor
a = exponent

The value of 'a' for many plants undergoing expansion is zero; there is a learning activity or improvement in the plant which offsets any increased workload due to production increase. The average value of exponent 'a' for all plants studied [10] averaged 0.18. Larger values of the exponent reflected parallel addition strategies or an increase in labour-intensive product handling. The relationship neglects radical organisational initiatives aimed at workforce reduction which have become more common in the 1990s.

Improvements in raw materials or energy consumption efficiencies depend largely on technological changes which are often introduced at expansions. Equivalent annual rates of improvement over plant life for all of the Australian plants studied averaged 0.6% for raw materials and 1.5% for energy per tonne of product. In some cases, however, improvements cannot be attributed solely to expansions, and in some cases, there are deteriorations in raw materials and energy efficiencies with increased plant production capacity. Corresponding equivalent annual rates of improvement in personnel and fixed capital per tonne of product for the plants studied averaged 4.5% and 3.0% per annum, respectively. Performance improvement in fixed cost areas was consistently greater than that in variable cost areas for all the plants studied.

As a result of plant expansion, one can confidently expect an improvement in personnel and fixed capital productivities, and if technological improvements can be incorporated, an improvement in raw materials and energy productivities also can be expected. The improvements, however, need to be compared with improvements realisable in new plants coming on stream incorporating best available technology and scale (referred to as 'BATS' plants).

Comparisons made between the expanding plants and new 'BATS' plants showed consistent improvement in the competitiveness of the expanding plants in fixed capital productivity but showed a consistent deterioration in competitiveness in personnel productivity and in some cases a deterioration in competitiveness in energy productivity. Productivity is defined here as the number of tonnes produced per unit of input—thus fixed capital productivity is the number of tonnes of product per real dollar of fixed capital investment, personnel productivity as the number of tonnes produced per person employed, and energy productivity as the number of tonnes of product produced per MWh (or alternative unit) of energy consumed. In terms of overall production cost, there was an estimated improvement in the competitiveness of the expanding plants, though performance on the expanding plant was still inferior to that on 'BATS' plants [11].

8.6.4 Factors contributing to noncompetitiveness in expanded plants

Factors contributing to noncompetitive performance in expanded plants include penalties arising from plant and equipment scale, process efficiency penalties from the older plant, and inferior safety and environmental standards.

Plant scale

Whilst an expanding plant grows in capacity during its operating life, there may concurrently be a more rapid growth in the capacity of new plants under construction. Growth in maximum available capacity for new plants was exponential and substantial in the 1960s for many chemical process technologies (13% per annum for ethylene, 15% per annum for ammonia, 17% per annum for ethylene oxide, 26% per annum for vinyl chloride monomer). Such growth rates greatly exceeded rates of capacity growth achievable in constrained expansions on operating plants over the same period. Thus plants, which are small in scale when initially constructed, can have their scale disadvantage exacerbated over plant life.

Equipment scale

Concurrent with exponential growth in capacity of entire plants, there has been an exponential growth in the maximum available capacity of equipment, notably process reactors. Such capacity growth has proceeded through the dual mechanisms of physical size increase and intensification.

Thus, where a plant expands by adding process reactors in parallel, it invariably inherits a greater number of reactors per unit of capacity than would exist on a new BATS plant. For example in 1983, the expanded ethylene plant referred to in Fig. 8.6 of 102,000 t/year capacity had nine cracking furnaces compared with a new replacement plant of 250,000 t/year capacity which had six cracking furnaces. Similarly, in 1983 the expanded aluminium smelter of 120,000 t/year had 544 pots compared with new smelters of the time of 240,000 t/year having 440 pots. These differences imply greater labour requirements per unit of capacity for both the operation and maintenance of the expanded plants.

Process efficiency

New plants generally offer higher efficiencies in raw materials and energy consumption per tonne of product than the expanded plants. This may be at the penalty of additional capital expenditures, but reflects the prevailing trends in unit costs of raw materials and energy and environmentally based objectives to reduce resource consumption and minimise greenhouse gas emissions.

Safety and environmental standards

It has been consistently observed for a range of technologies (for example in the manufacture of aluminium, chlorine, ethylene, vinyl chloride monomer, and soda ash) that there have been different environmental standards for new plants and older operating plants. Standards have been more stringent for new plants requiring additional capital expenditure on plant and equipment. Economic penalty is also likely to arise from protracted negotiations with planning authorities for project approval.

The use of parallel streaming (or series addition of equipment) in expansions tends to increase both the operational complexity of the plant and the number of potential emission points.

8.6.5 Risks and uncertainties in expanding existing plants

A number of uncertainties surround the expansion of operating plants. These are described as follows.

Extent of downtime to carry out plant modifications

This is a greater risk for single stream plants than parallel stream plants. Sequential modification to parallel reactors for example on parallel stream plants cushions the effect on plant downtime; there is also an opportunity for learning in that the last reactor modifications are usually made more quickly and cheaply than the first reactor modifications. Modifications requiring entire plant shutdowns can carry heavy financial penalties arising from lost production, especially if the shutdowns become unexpectedly protracted.

Determining the plant capacity distribution and potential for debottlenecking

This is more difficult for single stream plants with minimum intermediate storage. First-order bottlenecks to capacity (or those which restrict the increase of current production capacity) are easily identified. However, second- or third-order bottlenecks (which restrict further capacity development following the removal of first-order bottlenecks) are usually more difficult to establish.

Assessment of the ultimate life and capacity of the plant

Case study experience shows a consistent tendency to underestimate the ultimate life and capacity of plants, even after some years of experience in operating and expanding the plants concerned.

Ability to provide an adequate rate of capacity growth to match market growth

This will be a greater risk in the case of a constrained expansion, where the potential rate and extent of capacity growth are more restricted than for a parallel steamed expansion.

8.6.6 Optimum capacity of a new plant

We now return to the problem posed in Fig. 8.1 and consider the case of a new plant to provide additional capacity for a growing market. If the plant is built too small, its scale economies will be limited and it will also be short of capacity after a short period of market growth. If, however, the plant is built too large, it will suffer from poor capacity utilisation for too long in its initial operating life.

In selecting a suitable criterion by which to judge the best plant size, we need both a sound financial criterion and the facility to analyse why one plant size is preferred. The net present value (NPV), based on the forecast cash flows discounted at an interest rate reflecting the cost of capital, is the criterion preferred by most practitioners. Thus one looks for a peak in the profile of NPV versus plant size (see Fig. 8.7).

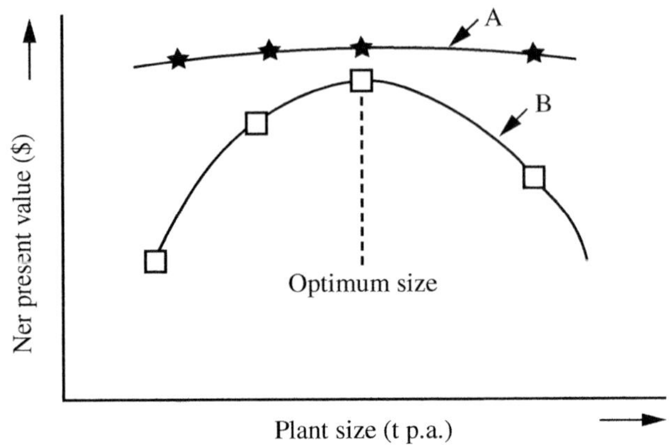

FIG. 8.7

Examples of possible NPV versus plant size profiles. (Either curve type may result from a cash flow analysis, depending on values of input parameters.)

For a given demand curve, the decision to build a larger plant may be interpreted as the decision to increase capital expenditure and certain fixed operating costs in the years prior to achieving full plant loading. This will be rewarded by greater sales and income in the years of the project following full plant loading (see Figs 8.8 and 8.9). Thus the decision to build a larger plant is inherently a long-term decision.

The additive property of the NPV is useful for sizing options where intermediate and derivative plants or associated utilities are involved. The net present value for total investment is then the sum of NPVs for the component plants.

Different size plants characteristically have a different shaped curve of NPV versus discount rate; in comparing two differently sized plants, the smaller plant may have a higher DCF rate of return but a lower NPV at the cost of capital (see Fig. 5.4 in Chapter 5). Some caution must therefore be exercised in the use of DCFR as a criterion.

The shape of the NPV versus plant size profile will depend on a number of factors:

- the demand curve for the product
- the effect of plant size on both capital and operating costs
- the discount rate
- project life
- the relative movements of costs and prices with time
- the capacity profile for the plant

Some of these influences with the aid of a worked example are discussed by Twaddle and Malloy [12]. Lower discount rates favour larger plants, since the cash flows resulting from sales at full plant output are not heavily discounted in relation to investment costs. Increasing project life also favours the larger plant, since the

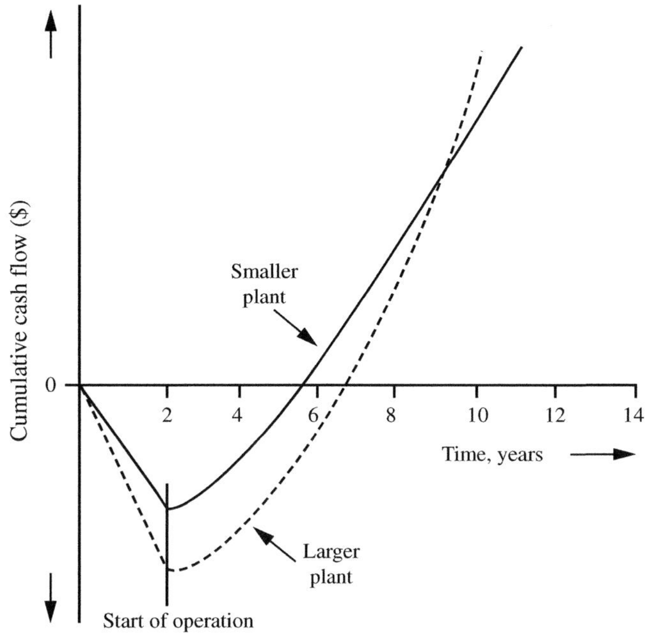

FIG. 8.8

Cumulative cash flow diagrams for plants of different size.

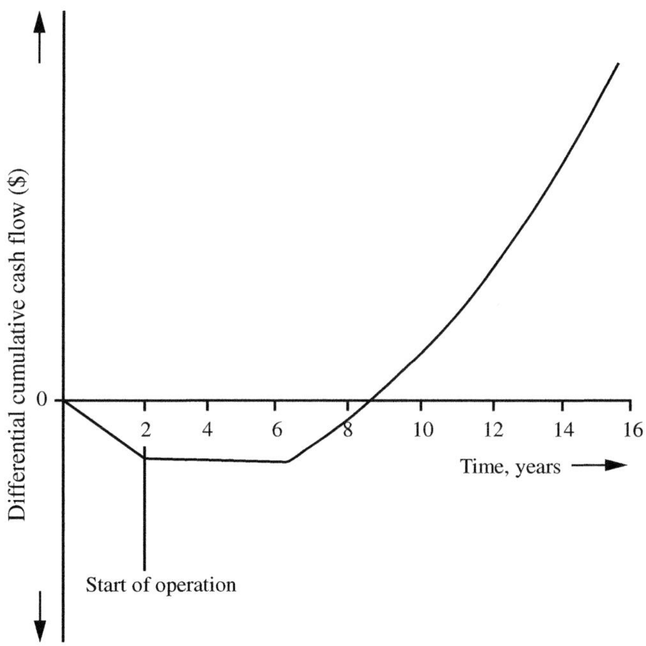

FIG. 8.9

Differential cash flow (cumulative) for larger versus smaller plant.

benefits are apparently prolonged. A slower escalation rate in selling price than in operating cost will result in a decreasing profit margin with time, thus favouring the smaller plant. The potential for capacity increase over the life of a plant will also favour the smaller plant option.

The interaction of these factors becomes complex, and since most of the longer-term projections are inherently uncertain, the plant sizing decision is one of judgement based on strategic considerations, rather than a mathematical optimisation. Nevertheless, the cash flow projections and NPV calculations are an important quantitative base for decision-making. The input data, being estimates and projections, can be subject to personal bias, as discussed in Section 5.9. This represents a challenge to project managers and team members to strive for objectivity.

8.6.7 Designing new plants with capability of future expansion

While designers may use judgement to anticipate future rates of technological change, the exact nature of technological change and its impact on their designs is uncertain. Plant design is also shaped by an operational philosophy geared to modes of raw materials supply and product despatch, to a particular cost structure, and to a particular market structure, all of which may change with time.

Because of uncertainties in forecasting and the penalties of higher capital expenditure, the provision of inbuilt capacity or capability for future expansion in a new plant should be examined critically. There are, however, specific provisions, which should be considered [7].

Provision of adequate space

- to allow for future installation of equipment, added in parallel or series
- to provide access for removal of obsolete equipment or modification of installed equipment

Such provision may incur additional capital initially but if integrated with safety and operability objectives, is likely to prove worthwhile. It assists future adding, replacing, or modifying equipment, without preempting the precise nature of change.

Provision of access points for measurement

This can be of great benefit in determining capacity distributions during the operational phase.

Judicial provision of capacity margins in selected equipment items

This is most appropriate for equipment items, which are initially expensive, where the subsequent capacity increase is likely to prove difficult and where the incremental capital required to provide initial capacity margins is minimal.

8.6.8 Choosing between a new plant and expansion of an existing plant

In choosing between the options of a new plant and expansion of an existing plant, considerations of long-term strategic objectives, current corporate position, and relative risks are important. A much larger capital investment is required for a new plant than for expanding the existing plant, carrying a higher risk in terms of potential economic failure. Technical and operational risks in expanding an existing plant have been outlined in Section 8.6.5, and these risks have their own economic implications. The operating company's corporate objectives at the time of decision could be (either individually or in combination) to:

- maximise cash flow
- maximise return on capital
- minimise production cost
- maximise sales volume
- conserve capital

An important financial aspect of a constrained expansion of an operating plant is the relatively minor impact on company cash flow arising from an investment compared with the case of a new plant of additional or replacement capacity. The lead time is normally longer for the new plant, reflecting the greater size and complexity of the project. Forecasting is necessary in relation to trends in technology and cost. Will investment in the new plant now preclude a better technological opportunity a few years into the future, for example? What is the forecast ultimate life and capacity of the existing plant and what will be its performance at that point? These questions need resolving since the alternative project concepts are so markedly different that it is difficult to apply conventional economic criteria in comparing the options.

8.7 Need for an industrial ecology—Capital, location, organisation implications

Many operating process plants operate within an industrial infrastructure. Friedlander [13] classified the scope of influence for process plants as:

- microscale activity for example an individual plant
- mesoscale activity for example petroleum refinery or petrochemical complex
- macroscale activity for example an industrial complex involving diverse industry sectors, such as power generation, chemical manufacture, and minerals processing

For a process plant involving the chemical transformation of raw material into a saleable product, the microscale activity is often difficult to implement as an economically viable option arising from the separate locations of raw material source and product demand. For example, a metallurgical smelter which processes mineral sulphides generates sulphur dioxide. The sulphur dioxide may be converted to sulphuric acid but there is then often the option of either transporting the moderately hazardous and relatively low-value acid product, or further processing of the acid into more valuable downstream products but necessitating further raw materials. The metallurgical smelter itself may be located at the minesite or require transportation of the mineral sulphides to a more suitable site for processing. Thus the interdependence of site location with feedstock availability and market for products, as well as other economic merits of the site influences process industry development and evolution.

At the same time, the land, capital, and industrial infrastructure requirements for a mesoscale activity are extensive, and a smooth, effective and economically viable operation takes time to evolve and is subject to changes in the market, technology, and financial influences.

Following the discovery of large reserves of natural gas off the North-west coast of Australia, the Australian government commissioned the Pilbara study [14] which examined the potential for a large-scale industrial complex in the Pilbara region. The advantage of large reserves of accessible natural gas, iron ore, and salt offered the potential for downstream processing arising from

- a large market for caustic soda in processing Australian deposits of bauxite to alumina
- opportunities to produce ethylene from ethane contained within natural gas, and hence further processing to ethylene derivatives
- opportunities to produce other derivatives from natural gas such as hydrogen, ammonia (and derivatives) as well as methanol (and derivatives)

However, there were a number of obstacles including

- relatively high capital costs and operating personnel costs arising from remoteness of the area
- the very large capital investment required derived from an extensive network of world scale plants necessary for competitiveness with global manufacturers
- the need to co-produce chlorine with caustic soda, demanding large-scale production of vinyl chloride monomer and PVC, for which there was only limited demand in Australia

Consequently, there have been large-scale exports of salt, liquefied natural gas, and iron ore for some 40 years with no downstream processing until the relatively recent construction of ammonia and ammonium nitrate plants in the region (Burrup Peninsula) derived from locally available natural gas.

8.8 Some cases involving challenges for sustainable investment decisions

The following cases involve integrating technical, economic, environmental, and sustainability considerations in making capital investment decisions. They are drawn from different Australian process industry contexts, but the considerations are applicable to decision-making in many countries.

8.8.1 Transition from mercury to membrane cell reactors in chlor-alkali plants

The chlor-alkali industry producing chlorine and caustic soda with small amounts of hydrogen from salt (as impure sodium chloride) is a major part of the established chemical industry, in United States and Europe, and many areas of the globe. Chlorine and caustic soda both have long histories of enabling processing into downstream products. The chlor-alkali industry has a long history and has undergone major technological change over its industrial life thus far. Some key developments include transition from carbon anodes to titanium anodes in mercury cells around the 1960s, and from mercury cells to membrane cells around the 1980s. The emergence of membrane cells was driven both by developments in membrane technology and an awareness of the environmental consequences of mercury emissions. Chlor-alkali production is energy-intensive, process providing further incentive for technology improvement in chlorine cells [15].

Chlorine and caustic soda production commenced in Australia from the 1940s with the largest production growth occurring in the Sydney plant at the Botany site where plant capacity for example grew from 8 kt/year in 1944 to around 95 kt/year in 1980 with progressive additions of new cell banks, introduction of technological change, and plant debottlenecking. A valuable insight into the economic, environmental, and safety considerations around the year 2000 in reshaping the chlor-alkali industry in Australia is provided by Clews [16]. Aspects discussed include

- the historical context of chlor-alkali production in Australia, dating from the 1940s, incorporating a growing response to market demands and an impressive growth in plant production capacity
- reductions in chlorine demand arising from the cessation of production of chlorinated solvents (carbon tetrachloride and perchloroethylene) and vinyl chloride monomer and PVC plants at the Botany site
- the locations, capacities, and other features of the nine chlor-alkali plants in operation around the year 2000
- detailing of the Australian markets for chlorine and caustic soda in the year 2000, and the imbalance in the relative demands for caustic soda and chlorine in Australia
- the challenge of replacing mercury cell plants with membrane cell plants, and consideration of design options for the new plants
- the cessation of the practice of liquefying chlorine at the Sydney site, due to the small distance between the new chlor-alkali plant and the surrounding urban population and related safety concerns
- technology selection decisions in relation to the choice of cells, other process decisions relating to brine purification and caustic soda evaporation, liquid chlorine storage at the Melbourne plant, and chlorine packaging

Drivers for the closure of the previously operating mercury cell plant and reinvestment included

- increasing fixed costs, especially high maintenance costs due to ageing equipment
- high personnel numbers required compared with modern plants
- concerns about mercury contamination of caustic soda and resulting market implications
- tightening regulation on mercury emissions to the environment
- high population density in close proximity of plant
- safety concerns about large inventories of liquid chlorine

Outcomes included

- a complete review by Orica of all its chlorine plants operating in Australia
- design and construction of two new membrane cell plants to replace obsolete mercury cell plants, with operations of new plants commencing in 2001
- elimination of chlorine liquefaction and liquid chlorine storage at the Botany site, with chlorine storage there reduced to low temperature, atmospheric pressure storage
- use of carbon dioxide in place of sodium carbonate, and use of a novel sulphate removal method, for brine treatment

8.8.2 Capture and utilisation of sulphur dioxide from metallurgical smelters

Australia is well endowed with a spectrum of mineral sulphide resources including lead, zinc, copper, and nickel. The capture and use of sulphur dioxide from metallurgical smelters treating mineral sulphide ores are important from both environmental and economic perspectives. Following the cleaning of gas leaving the smelter in order to remove particulates and any heavy metal impurities, process options include

- sulphuric acid manufacture involving oxidation of SO_2 to SO_3 followed by absorption of SO_3
- sulphur production involving reduction of SO_2 by methane
- gypsum production involving absorption of SO_2 in lime or lime/limestone slurries

Processing options are constrained by the required concentrations of SO_2 as inputs to the various processes

- an SO_2 concentration of 12% by volume is required for sulphuric acid manufacture
- an SO_2 concentration of >29% by volume is required for sulphur production
- an SO_2 concentration of <4% by volume is required for gypsum production

SO_2 concentrations leaving a metallurgical smelter vary depending on the mineral ore treated and the smelting technology used.

Sulphuric acid is the most common option adopted, but since sulphuric acid is seldom a marketable product at a remote minesite because of costs and safety of transport, it is also important to consider downstream processing options and transport to markets.

Main downstream product options are phosphoric acid requiring phosphate rock as raw material and phosphate fertiliser derivatives such as diammonium phosphate (also requiring ammonia as a feedstock) and superphosphate.

Estimates of costs and environmental impacts were made as part of a study [17] which explored processing options for smelter gases, and downstream processing of sulphuric acid into phosphate fertilisers in Australia. A number of minesites and coastal locations in Australia were considered. Fig. 8.10 shows the relationship between estimates of operating costs and environmental impacts for a range of processing options of smelter gases and downstream products where the smelter was located in a coastal city. The environmental index in this case was based on the life cycle assessment methodology and comprised a sum of normalised impact indicators for

- acidification potential
- global warming potential
- resource depletion
- solid waste generation

Details of the environmental assessment are reported in a separate paper [18].

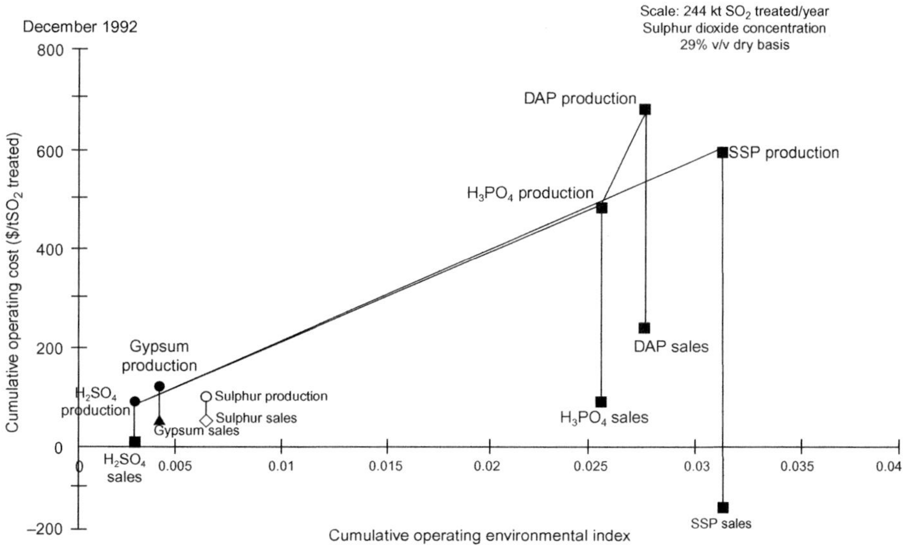

FIG. 8.10

Operating costs and environmental indices for SO_2 processing options [17].

8.8.3 IGCC with and without carbon capture

Integrated gasification combined cycle power generation (IGCC) has been proposed as a possible clean coal technology for reducing greenhouse emissions in coal-fired electricity generation. An assessment has been reported of the sustainability of IGCC technology compared with supercritical coal combustion for power generation from Queensland black coal [19]. The IGCC option was considered with, and without, carbon dioxide capture. The comparison was made using a coal consumption of 4369 t/day in each of the three cases. A feature of the study was the integration of technical, economic, environmental, safety, and sustainability criteria in assessing and comparing the cases.

The IGCC process with carbon capture was highly effective in reducing carbon dioxide emissions, with a reduction of 73% per MWh compared with the supercritical coal combustion case. However, there were economic penalties. The minimum viable selling price for electricity, assuming a 7% per annum cost of capital, was calculated ($US 2008) at

- $US80/MWh for the case of a new supercritical boiler with limestone desulphurisation of flue gas, but without carbon capture
- $US101/MWh for the case of IGCC with acid gas removal but without carbon capture
- $US 145/MWh for the case of IGCC with acid gas removal and carbon capture

There were environmental penalties for the IGCC cases, which whilst reducing SO_2 and NOx emissions, increased freshwater consumption by 2.5 t/MWh for both coastal and inland locations. There were also potential concerns regarding impacts of increased nitrate and sulphate concentrations in wastewater.

There were increased safety risks for the IGCC cases, with the Dow fire and explosion index assessed at 168 for the IGCC case, compared with 107 for the coal combustion case. The increased risk is related to the increased number of process units, and the severity of process conditions and gas compositions in many parts of the IGCC process. There were also risks of toxic gas release arising from high concentrations of CO and H_2S in the IGCC processing chain.

The study, reported in considerable detail, highlights the importance of integrating technical, economic, safety, and environmental assessments in weighing up the merits of alternative technology options.

8.8.4 Natural gas export from Australia's East Coast

Australia's domestic natural gas market has been briefly reviewed in Chapter 2. The Australian domestic supply is a composite of three segregated systems comprising the western, northern, and eastern segments. In the eastern market, pipelines connect South Australia, Victoria, Tasmania, New South Wales, and Queensland. The dominant domestic gas consumer in Australia has been the Eastern market with approximately 826 PJ per annum.

Perceptions of attractive global markets for LNG exports have encouraged exploration and investment in Australia into extraction, purification, liquefaction, and shipping facilities. Until 2018, all LNG exports from Australia had come from Western Australia and the Northern Territory.

Table 8.1 summarises some historical outputs in 2013 and projected outputs in 2018 for Australian natural gas consumed domestically and exported as LNG. The data is drawn from Australian government reports [20, 21].

In 2018 projected LNG exports of 1600 PJ resulted from three separate but coincident projects within the Easter market at Gladstone in Queensland. The export projects relied heavily on coal seam gas as the source of methane, with the following implications:

- much larger numbers of wells were required per tonne of gas than for conventional onshore sources
- underground saline water trapped within coal seams was extracted with the coal seam gas and required treatment
- potential conflict arose with land users and uses, particularly in relation to agriculture

The projected increases outlined in Table 8.1 indicate an equivalent growth rate in gas production for the Eastern market of 22% per annum over the 5 year period. This represents a rapid increase in production rate and has implications for learning in a number of contexts:

- those derived from the design and operating experience
- in relation to coal seam gas extraction implementation and related costs
- in dealing with environmental impacts from coal seam gas extraction
- in ensuring a balance in land use for resource extraction and agriculture
- in ensuring exports did not unduly restrict domestic use

The export projects, once operating, increased difficulties of supply and pricing for local users in the Eastern market. Price rises from \$4/GJ to \$9/GJ and beyond have been reported for local users.

Table 8.1 Historical and projected outputs for Australian natural gas.

Year		2013	2018
Western market	Domestic consumption	516 PJ	516 PJ
	LNG exports	1015 PJ	2435 PJ
Northern market	Domestic consumption	45 PJ	45 PJ
	LNG exports	249 PJ	605 PJ
Eastern market	Domestic consumption	826 PJ	826 PJ
	LNG exports	0 PJ	1600 PJ
Total	Domestic consumption	1387 PJ	1387 PJ
	LNG exports	1264 PJ	4640 PJ

When proposed, the Gladstone projects received enthusiastic support from the Queensland government [22] claiming benefits of 18,000 direct and indirect jobs, $3.2 billion increase in gross state product, $40 billion private sector investment, and royalty returns of $850 million per annum.

The difficulty of maintaining a balance between the supply of LNG to export markets and domestic gas consumers in Australia has been evident for some years. Western Australia formalised a reservation policy in 2006 ensuring 15% of the state's output from new LNG projects was to be reserved for sale in that state's domestic market. A gas reservation policy, however, has not been implemented on Australia's East Coast where most opponents recommend increased exploration and extraction of gas to solve the problem of meeting adequate domestic supply.

As gas exports from Gladstone have increased, difficulties for local users have intensified, leading to further attempts by the government to seek solutions. It is not unreasonable that potential benefits of the development were envisaged, but difficulties in maintaining current and future domestic gas supply and prices, risks associated with coal seam gas dependence, conflicts with landowners and agricultural initiatives, and the decision to proceed with the three LNG export projects in an instantaneous rather than a phased manner, raises a number of questions about the project's conception and implementation.

Some wider issues arising from this case include

- the ethical responsibilities for a country endowed with a resource in achieving the right balance between exporting the resource for the benefits of other countries and adding value within downstream industries in the home country
- accounting for environmental impacts associated with energy use and related greenhouse gas emissions derived from liquefaction, shipping, and regasification
- exploring the relative risks associated with the extraction of offshore and onshore gas (including coal seam gas)
- the long-term nature and commercial risks of gas extraction projects derived from different well head gas locations and compositions, government and environmental approvals, design and economic evaluations, and project approvals by the commercial operator and by government authorities

Related to the Gladstone LNG export case, it is useful to reflect on the importance of learning and rates of achievable rates in capacity growth discussed in Chapter 7. It is also of interest to reflect on Problem 6.6 within Chapter 6 dealing with phased investment. That case problem explored options for a natural gas treatment plant with a projected growth in throughput of 8% per annum. Findings were that whilst the option of matching plant capacity with forecast demand over project life involved capital and operating cost savings, a phased investment option had strategic advantages:

- future capital investment could be conserved if the projected demand growth failed to eventuate
- the second capital investment could reflect the design and operational learning from the first plant, as well as incorporating improvements in available technology

8.8.5 Regasification of LNG as a supply option for the East Coast

In response to the projected domestic shortages, proposals have been suggested in Australia during 2018 and 2019 from the local gas industry, for importing LNG from the West Coast or elsewhere to the East Coast for regasification.

Monash University chemical engineering students tackled this option as part of their design project in 2018, exploring effects of site location as well as regasification options, including the potential use of the cold energy derived from the storage of shipped LNG at −160°C at the regasification site. Other issues of importance derived from the environmental and safety implications of the shipping import terminal, unloading, storage, and regasification facilities.

One important aspect of economic viability in the project was the achievable capacity utilisation for shipping, storage, and regasification given seasonal influences of demand for domestic and commercial heating, as well as the risk of changes in demand in the wider market, and potential for new discoveries of onshore gas sources. A simplified estimate of operating costs showing the effect of capacity utilisation is shown in Table 8.2. Some allowance has been made in the estimate for reduced costs of utilities and reduced selling expenses for the 50% capacity utilisation case.

Table 8.2 Simplified, preliminary estimate of LNG regasification cost showing sensitivity to capacity utilisation.

Plant capacity	100 PJ/year		
Sales revenue	$900 million/year		
Fixed capital	$300 million		
Operating costs		Vaporisation cost $million/year	Vaporisation cost $/GJ
Electricity + other utilities		3.0	
Personnel employed including payroll overheads		5.0	
Maintenance materials		5.0	
Annualised capital cost		15.0	
Insurance		1.5	
Property taxes		1.5	
Fixed operating costs	100% capacity utilisation	28.0	
Selling costs		18.0	
Administration		1.0	
Total operating cost of vaporisation	**Based on 100% capacity utilisation**	50.0	**0.50**
	Based on 50% capacity utilisation	42.0	**0.84**

Experiencing a shortage of a key resource is a further reminder of the importance of resource depletion in environmental assessment methodology. One approach to assessing resource depletion is to explore the effects of substituting an alternative to the depleted resource. Rimos et al. [23] have explored the economic, environmental, and sustainability consequences of substituting black coal, coal seam gas, and diesel for natural gas in various industrial applications.

8.9 Reviewing completed projects

Process industry projects which have failed because of poor safety performance are much more widely and publicly scrutinised than projects which have failed on purely economic grounds. Major accidents such as Flixborough, Bhopal, Seveso, Piper Alpha, Longford, and Buncefield, referred to in Chapter 7, have led to legal investigations with a detailed examination of related causes, consequences, and liabilities. Published findings for the investigations have provided fruitful outcomes for learning about process safety, and about design and management of plant operations. This learning can then be effectively shared by industries, professional societies, and educational institutions, in the wider community.

Details of process industry projects which have failed economically are scarce in the public domain. Companies directly involved are reluctant to reveal details of their failures because of implications for their reputation and future business opportunities, which could extend to potential financers and business partners. Despite this lack of publicity, many projects fail to meet their economic objectives. Impressions are gained by outside observers of underlying causes, but these impressions are typically based on limited knowledge.

Some causes of failed projects are

- A failure of the market for the product to meet expectations. This might be due to the nature of the product, its lack of competitiveness, lack of demand, or other causes. There may also be contractual shortcomings in projected sales agreements between the supplier and purchaser(s).
- Excessive competition, where two companies which identify an investment opportunity which can strictly only support one project in the short term, nevertheless go ahead with investment and suffer poor capacity utilisation in the early years of operation due to market demand limitations.
- A failure of feedstock quality or availability to meet production requirements, especially in mineral processing projects.
- Limitations in the process technology adopted.
- Limitations in the skills or practice of the design, construction, and project management team.
- Inadequate time and detail allocated to project planning, evaluation, and approval.

- Unforeseen project delays caused by a variety of factors such as inclement weather at the construction site, limited skills or productivity in design, construction and project management personnel, delays in materials or equipment supply, delays or difficulties in plant commissioning.

Besides causing financial losses to the companies involved and their shareholders, failures can impact the morale of employees and wider stakeholders. Given the rigorous demands and uncertainties of process industry projects, it is important for companies to review all completed projects, however successful or unsuccessful, in order to learn from the strengths and weaknesses of the planning, evaluation, and implementation phases. Such learning should be documented, shared (at least with personnel within the companies concerned), and applied in subsequent projects.

An exception to the lack of publications on failed projects is an article by Rothman [24] focussing on failed projects in developing countries in the latter part of the 20th century. Many of the underlying causes reported by Rothman are related to inadequacies in the planning and management of projects. Besides identifying the underlying root causes of problems, Rothman provides a valuable list of recommended steps to avoid project failures.

Recommendation for further reading

The first reference is recommended for its thoroughness in assessing and integrating technical, economic, environmental and social criteria in process plant design. The second is listed because of its exploration of different perspectives in evaluating the economics of environmental improvement and regulation, supported by data.

1. A. Azapagic, A. Millington, A. Collett, Process design for sustainability: the case of vinyl chloride monomer, in: A. Azapagic, S. Perdan, R. Clift (Eds.), Sustainable Development in Practice, Wiley, 2004 (Chapter 6).
2. K.S. Rosselot, D.T. Allen, Environmental cost accounting, in: D.T. Allen, D.R. Shonnard (Eds.), Green Engineering. Environmentally Conscious Design of Chemical Processes, Prentice Hall, 2002 (Chapter 12).

References

[1] J.B. Guinee, Handbook on Life Cycle Assessment, Centre of Environmental Science/ Kluwer Academic Publishers, Leiden/Dordrecht, 2002.
[2] A. Azapagic, S. Perdan, R. Clift (Eds.), Sustainable Development in Practice, Wiley, Chichester, UK, 2004.
[3] D. Brennan, Sustainable Process Engineering, Pan Stanford, Singapore, 2012.
[4] G.A. Keoleian, D.V. Spitzley, Life cycle based sustainability metrics, in: M.A. Abraham (Ed.), Sustainability Science and Engineering. Defining Principles, Elsevier, Boston, USA, 2006 (Chapter 7).

[5] X. Xu, Economic Analysis of Landfills (PhD thesis), University of Queensland, Brisbane, Australia, 1998.
[6] T. Moilanen, C. Martin, Financial Evaluation of Environmental Investments, Institution of Chemical Engineers, Rugby, UK, 1996.
[7] D.J. Brennan, G.K. Stephens, Capacity expansion in process plants: some practical guidelines for successful strategies, Process Econ. Int. 7 (1988) 124–133.
[8] D.J. Brennan, G.K. Stephens, The extent and rate of capacity increase in process plants, Process Econ. Int. 5 (3 & 4) (1985) 15–25.
[9] J.B. Malloy, Risk analysis of chemical plants, Chem. Eng. Prog. 67 (10) (1971) 68–77.
[10] D.J. Brennan, G.K. Stephens, Changes in cost and performance resulting from plant expansion, Process Econ. Int. 6 (1) (1985) 16–26.
[11] D.J. Brennan, G.K. Stephens, Productivity trends in expanding plants and new plants, Technovation 6 (1987) 189–204.
[12] W.W. Twaddle, J.B. Malloy, Evaluating and sizing new chemical plants, Chem. Eng. Prog. 62 (7) (1966) 90–96.
[13] S.K. Friedlander, The two faces of technology: changing perspectives in design for environment, in: B.R. Allenby, D.J. Richards (Eds.), The Greening of Industrial Ecosystems, National Academy Press, Washington, DC, 1994, pp. 217–227.
[14] The Pilbara Study, Report on the industrial development of the Pilbara, June, 1974.
[15] I. Garcia-Herreo, et al., Life Cycle Assessment model for the chlor-alkali process: a comprehensive review of resources and available technologies, Sust. Prod. Consumpt. 12 (2017) 44–58.
[16] G.R. Clews, Replacement of mercury chlor-alkali plants with new membrane plants in Australia (Chapter 10), in: J. Moorhouse (Ed.), Modern Chlor-Alkali Technology, vol. 8, Blackwell Publishing, Abingdon, 2001.
[17] K.A. Golonka, D.J. Brennan, Costs and environmental impacts in pollutant treatment: a case study of sulphur dioxide emissions from metallurgical smelters, Process Saf. Environ. Prot. 75 (4) (1997) 232–244.
[18] K.A. Golonka, D.J. Brennan, Application of life cycle assessment to process selection for pollutant treatment, Trans. IChemE 74 (B2) (1996) 105–119.
[19] T.T.J. Falcke, A.F.A. Hoadley, D.J. Brennan, S.E. Sinclair, The sustainability of clean coal technology: IGCC with/without CCS, Process. Saf. Environ. Prot. 89 (2011) 41–52.
[20] Australian Government, Department of Industry Bureau of Resources and Energy Economics, Eastern Australia Domestic Gas Supply, Canberra, Australia, 2013.
[21] Australian Government, Department of Industry Bureau of Resources and Energy Economics, Energy in Australia, Canberra, Australia, 2014.
[22] Queenland Government, Queenland's LNG Industry. A Once in a Generation Opportunity for a Generation of Employment, November, 2010.
[23] S. Rimos, A.F.A. Hoadley, D.J. Brennan, Resource depletion impact assessment: impacts of a natural gas scarcity in Australia, Sust. Prod. Consumpt. 3 (2015) 45–58.
[24] S. Rothman, Industry can learn from past project failures, Oil Gas J. (2000) 37–40.

CHAPTER

Industry planning and structure

9

Industry...which dignifies the artist, lifts the swain,
And the straw cottage to a palace turns.
John Dyer (1700–58) Welsh clergyman and poet. From 'The Fleece' (1757)

Chapter outline

Process Industry Economics. https://doi.org/10.1016/B978-0-12-819466-9.00009-2

9.1 Industry classification

In considering process industries as distinct from process plants, the question arises—what constitutes an industry? The wider community tends to adopt a broad interpretation of industry as a sphere of work which generates income. Thus we have such diverse industries as the automobile, insurance, and entertainment industries. An industry represents an aggregate of activity, grouped according to the nature of resources used, products produced, or services rendered. The process industries, outlined in earlier chapters, may be broken down into a number of separate industries such as chemical, petroleum, metallurgical, food, pharmaceuticals, pulp and paper, fertilisers, plastics, and so on. Many of these are all part of a larger conglomerate, the manufacturing industries, though much of process industry activity is closer to resource-based industries such as petroleum and natural gas extraction, mineral extraction, and agriculture. The manufacturing industries can in turn be broken down into smaller components, with a hierarchy of aggregation adopted for classifying industries.

Most governments, through their relevant statistical authorities, have a system of industry classification, which provides a basis for collection and assembly of industrial statistics. These statistics, which cover aspects such as sales revenue, added value, employment, and extent of national ownership, assist in the assessment of contributions of industrial sectors to a national economy.

The Federal Government of the United States for example has three codes for classifying industries:

- Standard Industrial Classification (SIC)
- Enterprise Standard Industrial Classification (ESIC)
- Input–Output Classification Code

These are discussed in the context of the process industries by Couper et al. [1] in the book 'The Chemical Processing Industries Infrastructure', which draws on an earlier book by Wei et al. [2].

A detailed breakdown of SIC Group 28 for Chemical and Allied Products in the United States is outlined in Table 9.1. For comparison, the Australian and New Zealand Standard Industrial Classification (ANZSIC) System [3], introduced in 1993 and revised in 2006, covers 19 industrial divisions which include

- Agriculture
- Mining
- Manufacturing
- Electricity, gas, water, and waste services
- Construction

Table 9.1 Basic outline of major group 28 of the USA SIC.

Group no.	Industry no.	
281		**Industrial inorganic chemicals**
	2812	Alkalies and chlorine
	2813	Industrial gases
	2816	Inorganic pigments
	2819	Industrial inorganic chemicals, not elsewhere classified
282		**Plastics materials, synthetic resins, rubber, fibres**
	2821	Plastics materials, synthetic resins, and nonvulcanisable elastomers
	2822	Synthetic rubber (vulcanisable elastomers)
	2823	Cellulosic manmade fibres
	2824	Manmade organic fibres, except cellulosic
283		**Drugs**
	2833	Medicinal chemicals and botanical products
	2834	Pharmaceutical preparations
	2835	In vitro and in vivo diagnostic substances
	2836	Biological products, except diagnostic substances
284		**Soap, detergents, cleaning preparations; perfumes, cosmetics, and other toilet preparations**
	2841	Soap and other detergents, except specialty cleaners
	2842	Specialty cleaning, polishing and sanitation preparations
	2843	Surface active agents, finishing agents, sulfonated oils, assistants
	2844	Perfumes, cosmetics, and other toilet preparations
285		**Paints, varnishes, lacquers, enamels, and allied products**
	2851	Paints, varnishes, lacquers, enamels, and allied products
286		**Industrial organic chemicals**
	2861	Gum and wood chemicals
	2865	Cyclic organic crudes and intermediates, organic dyes and pigments
	2869	Industrial organic chemicals not elsewhere classified
287		**Agricultural chemicals**
	2873	Nitrogenous fertilisers
	2874	Phosphatic fertilisers
	2875	Fertilisers, mixing only
	2879	Pesticides and agricultural chemicals, not elsewhere classified
289		**Miscellaneous chemical products**
	2891	Adhesives and sealants
	2892	Explosives
	2893	Printing ink
	2895	Carbon black
	2899	Chemicals and chemical preparations, not elsewhere classified

Much of the Australian process industry activity is included within the manufacturing division comprising

- Food
- Pulp and paper products
- Chemical, petroleum, and coal products
- Nonmetallic mineral products
- Basic metal products
- Basic utilities supply
- Machinery and equipment
- Miscellaneous manufacturing

Within these subdivisions, industries are further classified into groups and in turn into classes. Table 9.2 lists some various groups and classes important to the process industries.

Table 9.2 Some groups and classes in Australian and New Zealand Standard Industrial Classification of relevance to the process industries.

Division	Group	Class	Title
A			**Agriculture, forestry, and fishing**
B			**Mining**
C			**Manufacturing**
	11		Food product manufacturing
	15		Pulp, paper and converted paper product manufacturing
	17		Petroleum and coal product manufacturing
		1701	Petroleum refining and petroleum fuel manufacturing
	18	1811	Basic chemical and chemical product manufacturing
			Industrial gas manufacturing
		1812	Basic organic chemical manufacturing
		1813	Basic inorganic chemical manufacturing
	182		Basic polymer manufacturing
	183		Fertiliser and pesticide manufacturing
	184		Pharmaceutical and medicinal product manufacturing
	19		Polymer product and rubber product manufacturing
	20		Nonmetallic mineral product manufacturing
	21		Primary metal and metal product manufacturing
D	26		Electricity supply Electricity generation
	27		Gas supply
	28		Water supply, sewerage and drainage services

9.2 Towards a taxonomy for the process industries

A suggested taxonomy for the process industries is now outlined. It comprises an ordered checklist which identifies key elements of an industry or its components requiring assessment. Some notes are added on many of these elements. The list can be further developed or simplified as appropriate for particular purposes. Many of the key elements deserve much closer consideration than they have received thus far in the chemical engineering literature. The checklist serves as a reminder of the complex characteristics of process industries and the frameworks within which they operate.

9.3 Suggested checklist for a taxonomy for the process industries

9.3.1 Markets

- local or export
- single or multiple products
- consumer good or producer good
- product classification: for example commodity, fine chemical, and speciality chemical
- safety and environmental characteristics of product in use
- market volume, locally and globally
- product selling price
- seasonal influences on demand
- product packaging
- product distribution
- impact of structure and future development of downstream industries
- competition from other products

Notes

1. These aspects are discussed in Chapter 2.

9.3.2 Feedstocks

- single or multiple
- source, reliability and pattern of supply, mode of supply
- transport and storage considerations
- cost
- classification for example naturally occurring, mined, hydrocarbon, agricultural, and processed
- impact of upstream industries and their future development
- demand and competition for feedstocks from other industries

Notes

1. These aspects are discussed briefly in Chapter 4.

9.3.3 Process route and process plant

- process technology adopted
- position within the overall manufacturing chain of related products
- maturity of process technology
- intensiveness regarding contributions of capital, feedstock, energy, and labour contributions to operating cost
- environmental and safety standards
- extent of local research and development capability to improve technology
- dedicated (single product) or flexible (multiple product and multiple grades of product) plant
- batch or continuous operation
- single stream or parallel stream
- extent of integration with other manufacturing plants
- age of plant
- single or multiple plants

Notes

1. Many of these aspects are discussed in Chapter 7.
2. Section 4.1 outlines the basis for cost intensiveness evaluation.
3. Flexible plants lend themselves to greater change and process development over operating life demanding a corresponding greater effort from technical personnel. Flexibility usually demands a higher initial and ongoing capital investment. Apportioning labour and capital-related costs between multiple products is more complex than with a single product.
4. Extent of parallel and single streaming influences economies of scale (discussed in Chapters 3, 4, and 7).
5. Age of plant will usually influence the extent of technological obsolescence though new technology can be introduced to older plants to improve performance and cost structure as discussed in Section 7.9.
6. Having multiple plants gives the operating company more flexibility, particularly if there are differences in location, feedstock, technology, and product properties for the various plants. Capacity utilisation can be varied over time on the different plants to gain cost or market advantages.
7. Remoteness of operations from R&D or technical activities (e.g. process design) places emphasis on communications. The operator needs both to identify and communicate technical needs to the R&D (or design) centre and to understand current developments and directions and their implications for operations. If technology is being purchased, there will be an additional business dimension to the interface.

9.3.4 Production

- capacity—maximum daily and equivalent availability
- capacity utilisation

- continuous or intermittent
- scale
- performance—productivity of labour, capital, energy, and feedstock
- production cost structure
- location—advantages and disadvantages
- constraints arising from raw material suppliers, market demand, and other influences

Notes

1. Capacity has been discussed under capacity planning in Section 8.6.
2. Intermittent operation may be necessary in extreme cases to accommodate available raw material supply and market demand patterns. Variations in supply/demand patterns result from seasonal effects in agriculture, for example milk supply and demand for fertiliser.
3. Scale effects are discussed under Sections 3.4, 4.15, 7.6, and 8.6.
4. Production cost structure is discussed in Chapter 4.

9.3.5 Possible government policy initiatives

- tariffs and other forms of import restriction
- investment incentive
- taxation
- royalties
- regulation of manufacturing and trading practices

Notes

1. Tariff is a duty charged by a national government on imported products to protect the local manufacture of those products. Most countries have adopted tariffs to protect industries in their early stages of development, but tariffs still persist in many countries for established industries. Tariffs are usually applied as a percentage of FOB prices in the country of origin (see Section 2.4). Import quotas may be applied to restrict quantities of imported goods independent of price. There is an international effort through the World Trade Organisation (WTO) to eliminate import restrictions and promote unrestricted trade between countries.
2. Investment incentives can take various forms such as investment grants, investment allowances (see Section 5.2), tax holidays (exemption from corporate tax for a number of years, usually to encourage new industries to a country or region), rapid tax depreciation (high rates of tax depreciation allowance), and so on.
3. Special tax incentives are sometimes offered to encourage research and development expenditure.
4. Governments frequently levy royalties on extracted resources (minerals or petroleum) and excise charges on consumer goods (beer or petrol) which affect downstream industries and consumers.

9.3.6 Financial performance at a company level

- capital formation and gearing at a company level
- capital investment—pattern and extent, policy, and financial capability
- profitability

Notes

1. Gearing or financial leverage is the proportion of a company's assets financed through debt or equity sources.

9.3.7 Administration

- ownership
- degree of autonomy in management decision-making
- organisational characteristics
- workforce
- extent of vertical and horizontal integration

Notes

1. Some of these aspects are discussed under Sections 9.4 and 9.5.

9.3.8 Impact on other industries and other sectors of the economy

- supplier of employment and tax revenue
- suppliers of raw materials, energy, utilities, maintenance services, consulting, and other services
- suppliers of technology
- user of products and other outputs from economy

Notes

1. Some of these aspects are discussed in Chapter 1.

9.3.9 National macroeconomic performance indicators

- gross domestic products of countries importing products
- gross domestic product of manufacturing countries
- patterns of international trade between countries for products of interest, including relative growth rates in supply and demand
- barriers to international industrial trade

9.4 Integration of process operations and industrial companies

Traditionally, many chemical engineers have been micro-oriented in their engineering approach to the process industries, typically working within a process plant, a design office, or a research organisation as a technical or technology specialist. It is important however to have a wider perspective of the process technology, plant and industry that one is directly engaged in, and its contribution to the economy. One important aspect of this wider perspective is transitioning from a process plant to an integrated processing site, such as a chemical processing complex or a petroleum refinery.

9.4.1 The chemical processing site

Increased emphasis on site-based performance has been occurring for several reasons:

- environmental effects such as noise and emissions to air impact at site boundaries irrespective of their source
- there are opportunities on a site to share effluent treatment facilities between plants
- there is an opportunity for exchange between process plants of surplus utilities and waste streams generated
- much related activity such as monitoring and reporting emissions, and dealing with government and community groups, is better managed under the one responsibility

Environmental pressures have been the driving force for product life cycle concepts, which require analysis of the entire manufacturing chain from a raw materials resource through to a fabricated or assembled product as part of the life cycle. This requires both a larger system for analysis and a broader systems perspective. The solutions to many waste treatment problems involve realising the waste as a useful by-product for raw material supply to a separate industry. Thus waste sulphur dioxide from metallurgical smelters can become a key raw material for the manufacture of sulphuric acid, which in turn can be used in the manufacture of downstream phosphate fertiliser. Linking apparently disparate industries through waste utilisation is one of the themes of the concept and practice of 'industrial ecology' [4].

9.4.2 The wider business and industrial context

It is important to know the technology framework for the process and product we are involved in, the strengths and weaknesses of related technologies, and the strengths and weaknesses of the relative manufacturing operations our company or organisation are involved in.

Knowledge of the wider process industries—their characteristics, structure and interaction—is an imperative to improved performance for a particular process industry, not only in the environmental context but in the context of overall economic performance. Other industries can be potential suppliers of raw materials and utilities, be potential markets for products, and use technologies and techniques which can be more universally applied. For example HAZOP studies, well established in the chemical industry, are equally applicable in the mining industry. Piping and instrumentation diagrams, well established in the chemical and petroleum refining industries, were later adopted by food industries.

The structure of a process industry can be defined, in part, according to the linkages between component plants, defining the material (and/or energy) flows between those plants in terms of mass and composition (and/or enthalpy and temperatures) to provide a processing network. The structure can further be defined according to the ownership, capital financing, and management of the component parts. Thus there are a number of ways in which industries and their components can be integrated.

Industry companies can integrate 'horizontally' (or 'laterally') by increasing the number of similar operations—for example the union of multiple ethylene manufacturing plants or the union of multiple dairy processing centres. The main purpose of horizontal integration is to achieve business economies of scale.

Companies can also integrate 'vertically' to become involved in upstream or downstream business activities. Vertical integration in manufacturing can be

- downstream (or forward) towards increased control of its markets
- upstream (backward) towards increased control of its raw materials suppliers

Thus an ethylene producer can integrate downstream by acquiring a vinyl chloride plant. A PVC manufacturing plant could integrate both downstream into PVC pipe fabrication and upstream into vinyl chloride manufacture.

Chemical and petrochemical industries are good examples of highly integrated industries. Salt is used as a feedstock for producing chlorine (and caustic soda), methane as a feedstock for producing methanol, and heavier hydrocarbons (ethane, propane, butane, etc.) used for producing ethylene. Chlorine, methanol and ethylene then become building blocks for downstream industries. Fig. 9.1 shows a diagrammatic representation of the Saudi Arabian petrochemical industry at an earlier stage of development [5] indicative of an industry which was then established, but also in the transition towards further development.

Integration of operations can occur on the one manufacturing site with considerable benefit. There are identifiable benefits in costs, and also safety and environmental risks where the product from plant A becomes the raw material for plant B where plants A and B are located on the one site. **Integrated manufacture** is to be distinguished from the case where plants A and B are on different sites requiring transportation of the intermediate from plant A to plant B. As the number of integrated plants on the one site is increased, so the potential economic benefits may also be expected to increase. Cost benefits arise from

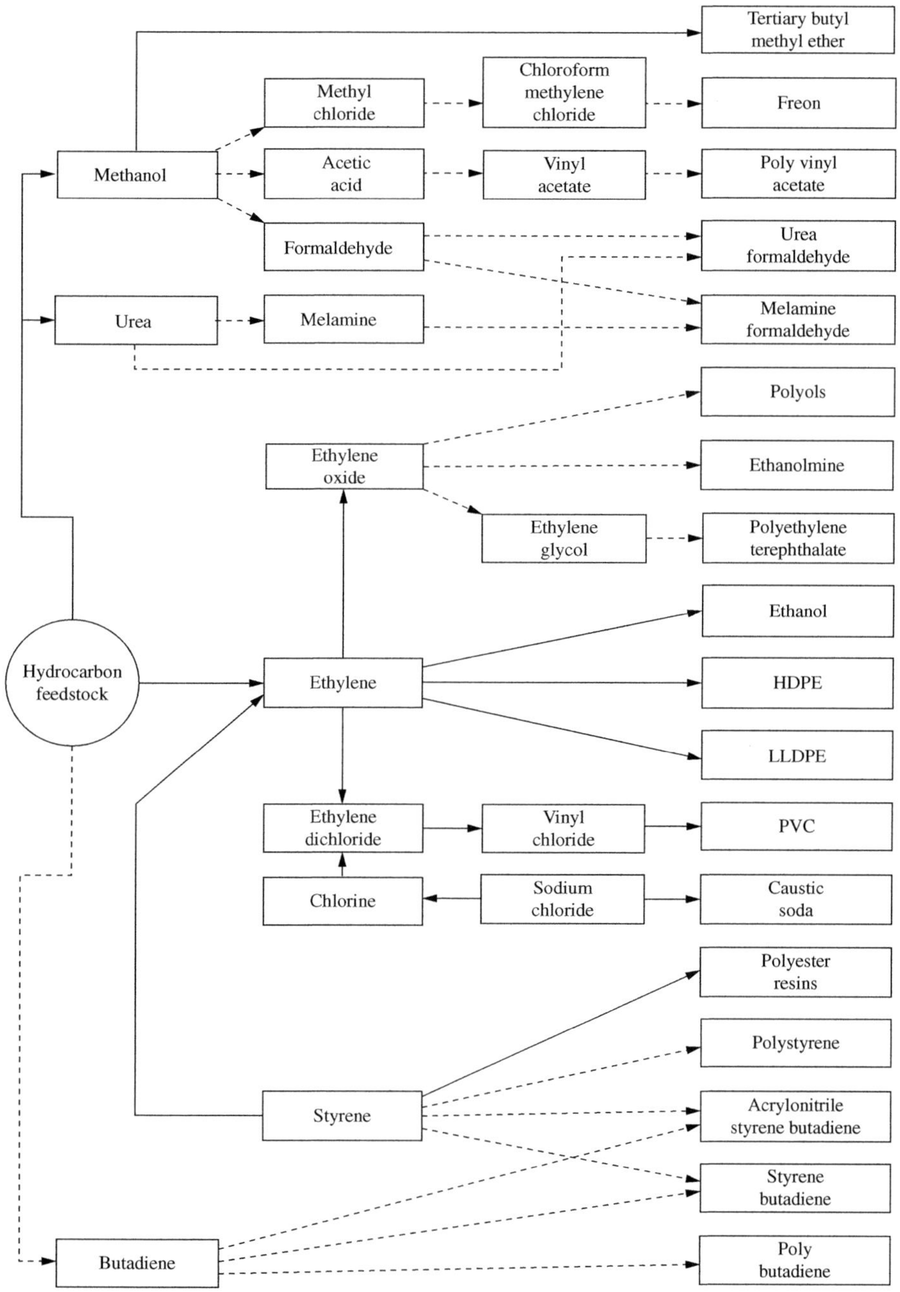

FIG. 9.1

Saudi Arabia Basic Industries Company at an earlier development stage. *Solid lines* denote established industries whilst *dotted lines* denote planned industries.

From K.M. Wagialla, Petrochemical feedstocks and downstream industrialisation, Process Econ. Int. 7 (1988) 74–91.

- reduction in outside battery limits fixed capital, due to reduction in storage requirements for intermediates, as well as scale economy benefits in shared utility facilities and buildings (see Section 3.3)
- reduction in working capital requirements due to reduced storage requirements for intermediates
- reduction in transport costs for intermediates
- potential for integrating energy requirements and surpluses, and using process streams which might otherwise be wasted
- available pools of skilled personnel on the one site

Safety and environmental benefits arise from

- confining transport of intermediates to within the site, usually using pipelines
- avoiding risks derived from rail, road, or pipeline transport in public spaces
- minimising risks in filling and emptying transport containers
- minimising risks derived from large inventories in storage vessels

Investigations by the author of Australian process industry sites indicate storage provisions of 15% of annual production for stand-alone cases and 2% of annual production for integrated cases.

The cost and safety benefits identified above must be weighed against some possible benefits of nonintegrated manufacture. Key raw materials or utilities may be available at a separate location. Integrated manufacture may impose increased reliability requirements and operational constraints for individual plants because of the high level of interdependence. Possible domino effects in disaster propagation from a single incident must be considered.

Integrated manufacture on the one site can involve a small number of plants, for example those producing ethylene, chlorine, vinyl chloride, PVC, and LDPE. A phosphate fertiliser industry could involve a number of plants on the one site linked as shown in Fig. 9.2. An example of integration at a high level is Ludwigshafen, BASF's global headquarters, where some 39,000 employees work on a site covering $10\,km^2$ and accommodating some 200 production units [6].

9.5 Location influences

In Chapter 3, some desirable characteristics of land and site location for process industry use were identified. Dominant commercial driving forces for location are access to raw materials and markets, but frequently only one of these criteria can be satisfied. There are industries which are resource located and those which are market located. For example there are petrochemical industries built up around the oil and gas supplies and salt deposits of the Middle East. Methane can be used to generate steam and electricity and to make hydrogen, ammonia, and methanol whilst ethane can be used as a feedstock for ethylene; both methane and ethane can be priced cheaply since they are by-products of a predominantly oil production industry

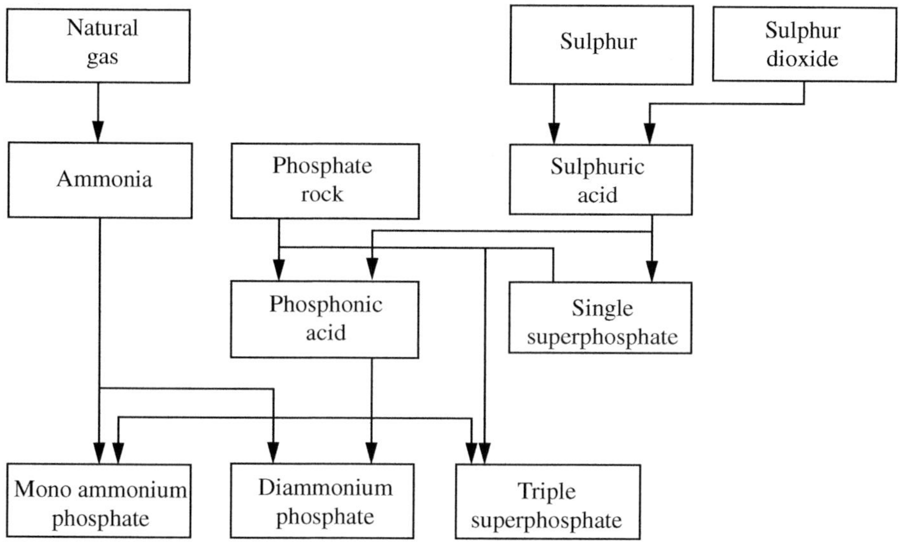

FIG. 9.2

Example of integrated manufacture of phosphate fertilisers.

and have often been flared in the past. Downstream products of a petrochemical industry can be made on a sufficiently large scale to enable a profitable export industry.

Likewise, the fertiliser industry can be usefully located near deposits of phosphate rock for example in North Carolina and Florida, United States, and Jordan in the Middle East. After mining and beneficiation, the phosphate rock can be processed on a large scale to phosphoric acid and downstream derivatives such as ammonium phosphates and triple superphosphate. These products are made on a large scale supported by large rock reserves and exported. Transport economics are more favourable for the higher value-added products.

Some national process industries, for example those of Japan, have very little indigenous raw material but have a large, sophisticated industrial market locally, supported by a large and confined population. Through innovation and locally developed technology, supported by good industrial planning and management, the disadvantage of higher raw material costs is offset to enable a highly competitive industry capable of servicing both local and export markets. Whilst raw materials must be imported thus incurring transport costs, being an important consumer Japan can source these in large quantities as a competitive and well-organised purchaser.

The benefits of integration have been discussed above in Section 9.3. Large, well-integrated sites can offer a new plant as well as the established plants cheap feedstock, cheap utilities, and other benefits derived from an established industrial infrastructure. Access to shipping through port facilities is a key location advantage and is usually part of the infrastructure of an established site. Examples of a chemical complex in Europe supported by large ports are those at Antwerp in Belgium, and Rotterdam in the Netherlands.

An example of an industrial complex in Australia is Kwinana, 40 km south of Perth in Western Australia. The complex was reported in 2008 as having an economic output of Aust$15 billion annually and employing 4800 personnel in core industries which included

- minerals processing and metal production
- petrochemicals and chemical products
- utility industries
- other manufacturing, transport, and service providers

A report published in 2008 [7] provides insights not only into the Kwinana complex, but how opportunities for synergies involving reuse of water, energy, and by-products have been explored, and the factors influencing success of related projects. The report also provides an understanding of the evaluation tools and organisational features employed in exploring and achieving various synergies.

The effect of location on plant capital costs has been discussed in Chapter 3. Remote sites incur capital cost penalties because of increased freight costs for equipment and materials to the site and the costs of attracting and accommodating the necessary workforce. Further cost penalties may be incurred due to geographical disadvantages associated with steep terrain or severe weather conditions. Many plants located at raw material sources suffer such isolation and geographical disadvantages.

Some sites are also more prone to environmental problems than others. This may be related for example to the sensitivity of the landscape (forest or desert) to certain emissions, or the availability or lack of space for solid waste disposal. There are also more stringent government environmental regulations in some countries than others as well as the effects of background pollution levels in certain regions.

The full spectrum of operating costs including transport also clearly depends on location. Location influences on operating labour and staff costs resemble those at project design and construction phase, referred to in Chapter 3. Raw materials and utilities costs can also be sensitive to location reflecting access, transport costs, government policy, scale, and other effects.

Superimposed on the natural advantages of industries and their national economies are the various government policies relating to business and industry. Many governments are prepared to offer tax concessions to encourage industry development. These are in the form of investment grants and allowances, tax depreciation allowances, tax holidays, tax concessions on research and development, and so on. A further consideration influencing the siting decision is the perceived and real political stability of a country.

Scale of plants is another industry characteristic which can be directly or indirectly related to location. Plant scale economies and decisions have been discussed in Chapters 3, 4, 7, and 8.

Ultimately a location decision for a plant is decided by weighing up many factors. Multiple objective analysis is a potentially useful tool for aiding decision-making in this context.

9.6 Industry ownership

The ownership of process industry companies within an industry is important in determining the location and hierarchy of decision-making and a whole range of policies on matters including capital investment, research and development, business specialisation and diversification, employment, company acquisitions and divestments. The proximity or remoteness (especially in the international context) of centres of research and development, or design or corporate management can have a major influence on the cultural development of a process industry business and the skills acquisition and career opportunities of its employees.

Multinational companies exercise considerable influence on the global allocation of material resources, personnel and capital and their respective impacts. This influence is exerted in the context of where the companies choose to acquire additional businesses, invest in new plants and locate their core activities such as design, research and development.

In process industry contracting, electronic communications now allow greater participation and exchange between countries in design and engineering services. This enables smoothing out of cycles in workload between different world offices and also enables wider use of less expensive engineering services in projects. This requires some uniformity in software, systems and standards between the various world offices.

Major changes in industry structure and orientation can result from amalgamation, divestment, and privatisation. These changes often impact on cost structures and prices (see e.g. Ref. [8] on trends in contract utility prices in Britain following privatisation of gas and electricity supply companies).

Detailed consideration of these topics is beyond the scope of this book but aspects are discussed in the more commercially oriented technical journals.

9.7 Industry organisations

Industry organisations are important vehicles for amalgamating the efforts of contributing companies into effective communication:

- between member companies
- with related professional societies
- with government members and government bodies
- with the wider community

Some important industry organisations in Australia include those focused on research and development like the Commonwealth Scientific and Industrial Research Organisation (CSIRO), whose research spans mining, mineral processing, chemical processing, and other process industry activities. There are also industry associations devoted to specific industries for example:

- Plastics and Chemicals Industry Association (PACIA) representing all sectors of the chemicals and plastics supply chain.
- Minerals Council of Australia, representing Australia's minerals exploration, mining, and minerals processing.
- The Australian Petroleum Production and Exploration Association, representing Australia's oil and gas exploration and production industry.
- The Australian Dairy Industry Council.
- The Australian Pulp and Paper Industry Technical Association representing the pulp and paper industry.

There are also important bodies related to energy and water supplies, for example:

- Australian Energy Market Operator, responsible for the administration and operation of the wholesale national electricity market in accordance with the National Electricity Code.
- Australian Water Association, Australia's leading water supply authority.

9.8 Professional societies

A number of professional societies are linked to the process industries including

- chemical engineering societies for example Institution of Chemical Engineers, American Institute of Chemical Engineers
- wider engineering bodies for example Engineers Australia, Royal Academy of Engineers
- chemistry professionals, for example Royal Australian Chemical Institute
- mining and metallurgical professionals, for example Australian Institute of Mining and metallurgy
- engineers and scientists working in the energy field for example Australian Institute of Energy

9.9 Government bodies

A range of government bodies is involved in diverse activities. In Australia, government bodies operate at commonwealth government, state government, and local councils in matters relating to energy supply, water supply, materials recycling, and resource management.

Examples of bodies in the United States include

- Environmental Legislation for example US Environment Protection Authority.
- Safety Legislation for example Occupational Safety and Health Association.
- Business policy for example Federal Trade Commission (FTC).

9.10 Chemical engineering contribution to government policy development

Chemical engineers through their understanding of the process industries and the technical, economic, environmental, and safety criteria are in a strong position to contribute to government policy development and related initiatives. Some areas relevant to broad government responsibilities impacting on the needs of both industry and the wider community include

- development of environmental regulations, including related economic incentives
- development of safety regulations
- provision of utilities particularly
  - Electricity
  - Water
  - Gas
- waste treatment and disposal
- materials recycling

Professional societies are increasingly recognising the importance of contributing to policy development by governments and also understanding the role of government and government departments. There are pressing needs for the chemical engineering community to be more active and effective in this area [9]. The Institution of Chemical Engineers (IChemE) has recently become more committed for example through its energy centre, and in policies related to safety, materials recycling, and the food chain [10]. The National Engineering Policy Centre, led by the Royal Academy of Engineering in the United Kingdom, is also aiming to bring engineering expertise to government policy development [11]. The policy centre includes some 38 organisations representing engineering, including IChemE.

9.11 Industry data sources

Data sources for market evaluation and capital and operating cost estimation have been identified in Chapters 2–4. Insights into policy and practice of various industry sectors and companies as well as much useful quantitative data can be gained from

- industry association publications
- government reports (particularly those emanating from government enquiries into particular industries or sectors)

- company reports
- the more commercially oriented technical journals

Annual company reports are a valuable source of financial information in terms of wealth creation, balance sheet, and profit and loss statements, for the company concerned. A company's financial records over recent years might also indicate its capacity to fund a new project or major expansion.

Three examples of useful information sources are now reviewed. Examples A and B provide valuable insights into trends, performance measures, and influences in the chemical industries in the United States, and Europe. Example C focuses on exported Australian resources, with details of global trends in supply and demand and selling prices for a wide range of energy and mineral resources, together with an analysis of influential macroeconomic factors.

9.11.1 Example A. 2018 Elements of the Business of Chemistry [12] published by the American Chemical Council

This comprehensive report provides an insight into a number of technical and financial aspects of the chemical industry, both in the United States and globally. The report also gives insight into technical details of the US chemical industry, a number of economic perspectives, and some valuable guides to information sources of related data.

Some features include

- **Maps of chemical chains** (provided in Appendix A) showing downstream derivatives from the following base chemicals:
  - Chlor-alkali
  - Methanol
  - Ammonia
  - Ethylene
  - Propylene
  - C4 hydrocarbons
  - Benzene
  - Toluene
  - Xylene
- **Comparison of relative cost structures** for the categories of basic chemicals, specialty chemicals, pharmaceuticals, and consumer products produced in the United States in terms of
  - Feedstock and other raw materials
  - Labour
  - Advertising
  - Other operating expenses
  - Environmental, health, and safety spending
  - Taxes
  - Profits

The data indicate for example a dominance of raw materials costs for basic and specialty chemicals, and higher profits for pharmaceuticals and consumer products.

- **Shipment values in US$** for basic chemicals, specialty chemicals, agricultural chemicals, pharmaceuticals, and consumer products. The data indicate that basic chemicals and pharmaceuticals are the major contributors. The term shipment is used in the report to mean the net selling values on a free on board (fob) basis to customers of products shipped from manufacturing facilities.
- **Global comparisons of chemical shipments, chemical imports and exports, and capital spending** for the six regions of North America, Latin America, Western Europe, Central and Eastern Europe, Africa and Middle East and Asia Pacific.
- **A comparison of total global chemical shipment values** for the categories of basic chemicals, specialty chemicals, agricultural chemicals, pharmaceuticals, and consumer products.
- **A comparison of value of shipments ($million), employment (numbers), average annual wages and salaries ($), and value of chemical exports ($million) for the States of United States.**

  The data shows the value of shipments greatest in Texas, California, North Carolina, and Louisiana, whilst average wages and salaries varied considerably between states.
- **Trade details including**
  - trading partners
  - trends over time in US trade balance
  - trends over time in the value of imports and exports to and from United States
  - share of imports and exports for product sectors (i.e. basic chemicals, speciality chemicals, agricultural chemicals pharmaceuticals, and consumer products)

Additional details of interest include US chemical industry performance in relation to

- **Environmental, health, safety, and security spending**
- **Emission indicators over time from 1985 to 2017 for**
  - Sulphur dioxide
  - Nitrogen oxides
  - Volatile organic compounds
  - Carbon monoxide
  - Coarse and fine particulates
  - Ammonia
- **Greenhouse gas emissions from 1990 to 2017, reported both as direct and indirect emissions**
- **Worker health and safety statistics**
- **Pattern of energy consumption in different segments of the chemical industry including**

- share of total energy consumption by source (e.g. coal, natural gas, electricity, etc.)
- trends in energy efficiency over time

- **Transportation quantities of products by mode (i.e. truck, rail, waterborne, pipeline) and transportation costs by mode for chemical products**

9.11.2 Example B. Facts and Figures 2018 of the European Chemical Industry [13] published by CEFIC, the European Chemical Industry Council

The CEFIC report includes

- global comparisons of chemical sales by country
- breakdown in European chemical sales value by-product sector (i.e. petrochemicals, specialty chemicals, basic inorganics, polymers, and consumer chemicals)
- breakdown in chemical industry sales from the European Union (EU) by geographical region
- trends in chemical industry sales revenue from the EU over time from 2006 to 2016 in terms of
  - total sales
  - home sales, that is where products are manufactured and sold within the one country from the EU
  - intra EU sales where one country within the EU sells products to another country within the EU
  - extra EU sales, where products manufactured in the EU are sold to countries outside the EU. These represented around 30% of total sales revenue in 2016
- price comparisons for electricity and gas with other countries
- trends in energy consumption over time and patterns of consumption for different fuel sources
- capital spending in the EU over the period 2006 to 2016, and global comparison of capital spending in the EU in 2006 and 2016
- research and development spending in the EU from 1996 to 2016, and global comparison of R&D spending in 2006 and 2016
- impacts of regulatory costs on profitability in EU chemicals sector

9.11.3 Example C. Resources and Energy Quarterly [14] published every 3 months by the Department of Industry, Innovation, and Science, within the Commonwealth Government of Australia

The report provides an overview of macroeconomic performance of importing countries and international trade arrangements influencing exports of mineral commodities from Australia, and their contribution to Australia's economic growth. World consumption and production patterns, as well as selling price trends, are reported for commodities including iron ore, thermal coal and metallurgical coal, natural gas, uranium, gold, aluminium, copper, nickel, zinc, and lithium. The report provides information on upstream processing of minerals within Australia, usage patterns for the commodities, trends in Australian production and exports, and trends in international selling prices.

9.12 Human resource contributions

It is important to reflect on the contributions of human resources in the sharing and development of expertise in process economics. Expertise is owned by individuals and groups within

- companies involved in manufacturing
- design and project management companies
- consulting companies
- bodies involved in research and development
- industry associations
- professional societies
- government bodies
- academic institutions

In Chapters 2–5, key contributions to the economic evaluation of projects were explored from

- market forecasting
- capital cost estimation
- operating cost estimation
- profitability evaluation

Frequently these tasks are conducted separately by specialists and their organisations, building on established expertise and prior experience, supported by the accumulated data gleaned from earlier projects or studies. Successful investments depend on the ability to bring the contributions of such specialists together.

In Chapters 7 and 8, the importance of environmental, safety and sustainability criteria, and their interaction with technical details and economics was explored. Implications for time-dependent learning and implementation in investment decisions were explored with some case examples. Interaction of experts in these areas with those doing economic evaluations is likewise an important requirement for success in project evaluation and implementation.

Governments and their departments have an important role in developing and defining policies related to economic success in the process industries, in encouraging technology development, in encouraging productive investment, in safety and environmental regulation, and in ensuring that the needs of stakeholders, including the wider community, are nurtured.

Education institutions have a key role in providing foundation knowledge and skills and encouraging dialogue between researchers, educators, government departments, professional societies, and industry organisations to explore opportunities and work towards long-term benefits derived from process industry investment and activity.

A culture of working towards skills development and towards a sharing of knowledge and skills, both within and across the organisations identified above, is a major consideration in assisting process industry improvement and sound economic performance.

References

[1] J.R. Couper, O.T. Beasley, W.R. Penney, The Chemical Process Industries Infrastructure. Function and Economics, Marcel Dekker, New York, USA, 2001.

[2] J. Wei, T.W.F. Russell, M.W. Swartzlander, The Structure of the Chemical Processing Industries: Function and Economics, McGraw-Hill, New York, USA, 1979.

[3] Australian Bureau of Statistics ANZSIC Industrial Classification, 1292.0, September, 2008.

[4] B.R. Allenby, D.J. Richards, The Greening of Industrial Ecosystems, National Academy Press, Washington, DC, USA, 1994.

[5] K.M. Wagialla, Petrochemical feedstocks and downstream industrialisation, Process Econ. Int. 7 (1988) 74–91.

[6] https://www.basf.com/global/en/who-we-are/organization/.../ludwigshafen.html (Last accessed August 2019).

[7] Centre for Sustainable Resource Processing, Capturing Regional Synergies in the Kwinna Industrial Area, 2008 Status Report, July, 2008.

[8] The utility market—where do we go from what we have learnt?, Energy World (September) (1996) 14–15.

[9] D. Brennan, Developing policies: what can we do? Increasing chemical engineering contributions to sustainable policy development: an Australian perspective, The Chemical Engineer, (November), 2018, 48–51.

[10] N. Choudhury, Policy perspective, The Chemical Engineer (September) (2019) 55.

[11] N. Starkey, Policy central, The Chemical Engineer (September) (2019) 53–54.

[12] American Chemical Industry Council, 2018 Elements of the Business of Chemistry, (July 2017).

[13] CEFIC, Facts and Figures 2018 of the European Chemical Industry, (December 2018).

[14] Australian Department of Industry, Innovation and science, Resources and Energy Quarterly (June) (2019).

Glossary

Availability The extent to which plant or equipment items are 'online' or 'available' for use in production, usually expressed either fractionally or in days per year. 'Equivalent availability' is the equivalent number of days per year that a plant can produce at maximum daily capacity. Equivalent availability accounts for periods of unavailability (or shutdown) of the entire plant, and also for periods where production is restricted by internal factors (such as fouling, wear, loss of control, unavailability of certain equipment items).

BATS plant New process plants of best available technology and scale.

BAT plant New process plants of best available technology but smaller than the largest available scale.

Battery limits A geographic boundary around the equipment items which collectively constitute the process plant. Plant facilities outside the boundary are termed *offsites* or *outside battery limits*.

Campaign A production run of significant duration punctuated by the shutdown for plant cleaning, maintenance or modifications. Successive campaigns may be made for a single product or for different products.

Capacity (of plant) The maximum production capability of a process plant per unit time. 'Maximum daily capacity' expressed in tonnes/day (or similar units) is the maximum production capability in a short time span, usually assessed by a high production rate trial over a period of 1–3 days. 'Annual capacity' is the product of maximum daily capacity and the equivalent availability of the plant to produce at maximum daily capacity.

Capacity utilisation The ratio of production achieved over a given time span (adjusted to a daily equivalent) to maximum daily capacity.

Capital recovery factor Factor of capital investment required to provide equal annual repayments of the investment with interest over a given time period.

CIF (cost, insurance, and freight) A trade arrangement where the exporter pays the cost of supplying a consumer with a product, inclusive of all freight and distribution costs, including insurance.

Circular economy The concept and practice of enabling recycling of surplus materials over the manufacture and use of a product to reduce waste disposal and reduce resource consumption.

Commodity A product traded internationally and essentially to the same specification.

Constrained expansion A plant expansion where parallel addition of equipment is precluded in at least one step within the processing sequence.

Consumer good A product which requires no further processing prior to use by the ultimate consumer.

Contingency Allowance in cost estimates for unforeseen cost elements which are likely to occur. These include costs resulting from minor design changes or from unforeseen changes in the conditions under which the project is undertaken such as weather, currency exchange rate, or inflation.

Cost of capital The weighted cost of money from all sources, equity and loan, required to fund a project, expressed as an interest rate, typically per cent per annum.

Creditors Persons or companies to whom money is owed by reference company, for example for purchase of raw materials or utilities.

Debottlenecking Utilisation of potential excess capacity in some sections of a plant by eliminating capacity restrictions in other sections.

Debtors Persons or companies owing money to reference company, for example for sale of products.

Depreciation The decline in value of an asset (typically plant or equipment) over its life. Book depreciation allowance is an internal company allowance or transfer set aside to fund an eventual replacement. Tax depreciation allowance is an allowable deduction from income for assessing corporate tax payments.

Discounted cash flow rate of return The interest rate at which the net present value for a project is equal to zero. The discounted cash flow rate of return must exceed the cost of capital for a project to be profitable.

Elasticity (price/demand) The dependence of product selling price on demand for the product.

End use analysis A method for estimating market volume for a product in which product end uses are identified and projected.

Equity The funds owned by a company, including those from new share issues and retained earnings from profitable operation.

Exchange rate (or currency exchange rate) The rate at which one international currency can be exchanged for a currency from another country.

External cost or externality A cost that is outside the scope of a cost estimate but has an external influence on the environment or society. A common example is the cost of environmental damage experienced as a result of emissions from a power station or an industrial process plant.

Fine chemical A chemical made in smaller volume and to tighter product specifications than a commodity.

Fixed capital Capital investment into plant and buildings to enable product manufacture. This investment is essentially nonrecoverable.

Fixed operating cost A cost, which when expressed in dollars per unit of time, is unchanged with a change in production rate.

Floor price A minimum selling price for a product required to make a minimal return on capital investment when using competitive scale, technology, and unit costs.

FOB ('free on board') price Price of a product or an equipment item on board ship in the country of origin (or at the manufacturer's location). The purchaser is responsible for the costs of transport and related insurance to the point of delivery.

Grass roots plant A complete plant constructed on a virgin site.

HAZOP Hazard and operability (study) used to explore the effects of deviation from design intent for process parameters such as flowrates, pressures, temperatures on plant performance, particularly in regard to safety and operability.

Intensification An increase in production capacity per unit of equipment size. In new plants, this can be recognised as a decrease in the size of equipment for a given process duty. Process intensification is frequently achieved by improvements in reactor yield or selectivity, by conversion from batch to continuous operation, or in batch processes by shortening batch times. Benefits of intensification include reduced capital costs and process material inventories per unit of production capacity.

Intermediate A term used to denote a product which is 'intermediate' between a basic feedstock and an end product in the industrial processing sequence. For example, ethylene is an intermediate between ethane and polyethylene.

Internal rate of return A term having the same meaning as discounted cash flow rate of return.

LCA (life cycle assessment) A method of environmental assessment which provides a quantitative assessment of the environmental impact of a product over its life cycle. It includes an inventory analysis defining material and energy inputs and outputs, followed by classification, evaluation, and improvement analysis phases. LCA can also be applied to segments of the life cycle, for example, in evaluating the environmental impact of a chemical process.

Learning curve Term used (often synonymously with 'experience curve') for improvement in performance or reduction in production cost (or selling price) with increased cumulative production.

Location factor A factor which expresses the relative fixed capital cost of a plant in a given location compared with that in a reference location.

Marginal cost (also termed 'incremental cost') The production cost per unit of product for an increment of production beyond a given production rate. The term margin is also used in a pricing context as a margin above the floor price.

Net cash position Equivalent to the cumulative cash flow at any stage of project life, it can be read directly from the cumulative cash flow diagram for the project.

Net present value The net sum of the present values of all cash flows incurred over the life of a project, incorporating investment, operational and termination phases.

Nominal dollars (or alternative money units) Dollars in the year of expenditure without adjustment for inflation.

Offsites A term used to denote 'outside battery limits' facilities including storages (for raw materials and products), utilities, and facilities such as offices, workshops, laboratories, cafeteria, medical units, and other similar facilities.

Pallet A moveable platform for the storage of packaged products and their transfer by forklift truck.

Parallel streamed expansion Plant expansion involving the addition of an entire new process stream, or the parallel addition of equipment as required at a given step within the processing sequence.

Patent The legal right given to an inventor of process technology to control the use and practice of the invention. One right is to be the sole user of the invention for a nominated period, usually 16 years.

Payback time A term used to denote the time taken in a project to recover investment costs. It is sometimes taken as the time from commencement of the project to recover the total capital investment. More frequently, it is taken as the time from the start of production to recover fixed capital expenditure.

Product life cycle A term used to describe the complete scope of extracting the required raw materials, the various processing stages required to produce the product, the use of the product, and the ultimate disposal of the product.

Present value The present value of a future cash flow is the value of cash which, if invested now, would grow to the same value as the future cash flow due to compounding interest.

Producer good An industrial product which must undergo further processing before reaching the consumer.

Productivity The number of tonnes of product produced per unit of input. For example Raw materials productivity: Number of tonnes of product per tonne of raw material Energy productivity: Number of tonnes of product per MWh (or alternative unit) of energy Personnel productivity: Number of tonnes of product per annum per person employed Fixed capital productivity: Number of tonnes of product per annum per real dollar of fixed capital The term 'labour productivity' can similarly be used in relation to the number of workers required to achieve a particular goal in a construction, operation, or maintenance activity

Real dollars (or alternative money units) Dollars adjusted for inflation to a common time basis using an appropriate deflator. The consumer price index is most widely used as a deflator of prices, costs, or cash flows. Other indices may be more appropriate for specific purposes, for example plant cost inflation indices for examining trends in plant costs over time.

Revamping Physical modification of a plant or equipment item leading to a change in hardware or the mode of operation. Common reasons for revamping include increase in plant capacity, reduction in production cost, accommodating to change in feedstock or product composition, and in meeting more stringent environmental or safety standards. Revamping frequently involves the introduction of new technology.

Royalty Payment for the provision of technology. Payment may be made to a technology licensor as an initial lump sum or an ongoing payment based on sales (or a combination of the two). Royalty payments may entitle the purchaser to 'know-how' embodied in new plants, and to subsequent improvements in technology which can be incorporated into operations or plant modifications.

Scale Strictly this term means the size of an equipment item of specific geometry and design. It has been more widely (and perhaps loosely) employed to incorporate the concept of plant production capacity, where the same process technology is employed.

Speciality chemical A chemical product developed to meet specific customer needs and identified by function rather than its correct chemical name.

Spot price Selling price for a product outside of contracted arrangements and typically for limited volumes over short time frames.

Unit cost A cost per unit of raw material, energy, utility, labour, or other input. Examples include the cost of steam in $/tonne, cost of ethylene in $/tonne.

Variable operating cost A cost, which when expressed in dollars per unit of time, varies with production rate. Common examples include the costs of raw materials and utilities.

Working capital Capital investment over and above land capital and fixed capital required to initiate and sustain the operation of a process plant. Working capital includes stocks of raw materials and finished products, and extended credit (or 'debtors' minus 'creditors').

APPENDIX

Some currency exchange rates relative to the US$ for selected countries

1

		June_1990	June_1997	June_2010	June_2015	June_2019
Country	**Currency**					
United States	US$	1	1	1	1	1
Australia	Aust $	1.27	1.34	1.17	1.3	1.43
Canada	Can $	1.16	1.3	1.05	1.24	1.31
Europe	Euro			0.819	0.894	0.89
Japan	Yen	128	115	88.6	122	108
New Zealand	NZ $	1.7	1.47	1.44	1.47	1.49
Taiwan	Taiwan $	27.2	27.8	32.2	30.9	31
United Kingdom	Pound	0.572	0.6	0.572	0.635	0.792

Notes: *The above data for sample countries has been taken from International Monetary Fund (IMF) published statistics. Extensive data are available for a wide range of countries for each month of current and previous years.*

2 Consumer price index (CPI) statistics for some selected countries

Year	1995	2005	2010	2012	2014	2015	2016	2017	2018
Country									
Australia	67.6	86.3	100	105.1	110.4	112	113.5	115.7	117.9
Canada	75.2	91.9	100	104.5	107.5	108.7	110.2	112	114.5
China			100	108.2	113.2				
China						100	102.0	103.6	105.8
India	37.6	65.8	100	119.0	140.4	148.6	155.9	159.8	167.6
Germany	80.5	92.5	100	104.1	106.6	106.9	107.4	109.3	111.4
Japan	101.1	100.4	100	99.7	102.8	103.6	103.5	104.0	105.0
South Africa	43.4	74.3	100	110.9	124.4	130.1	138.4	145.7	152.4
United Kingdom		87.4	100	107.4	111.8	111.9	112.6	115.6	118.4
United States	69.9	89.6	100	105.3	108.6	108.7	110.1	112.4	115.2

The statistics are available for a wide range of countries over and extensive time record.
Comments: *As an example, the CPI is measured in Australia as a basket of goods and services representative of those required by private metropolitan households. The 'basket' can be subdivided into 11 major groups: food and nonalcoholic beverages; alcohol and tobacco; clothing and footwear; housing; furnishings, household equipment, and services; health; transport; communication; recreation and culture; education; insurance and financial services.*
Source: United Nations Monthly Bulletin of Statistics Issue 1178, August 2019.

Capital cost inflation index for Australian plants cost index

3

Index estimated by the author based on data published by the Australian Bureau of Statistics.

Financial year ending June	Materials price index	Producer price index	Average weekly earnings ($)	Plant cost index	Consumer price index
1981					100.0
1982	125.4		283.75	100.0	110.4
1983	139.6		324.15	112.8	123.1
1984	147.3		349.45	120.3	131.6
1985	155.5		376.08	128.2	137.2
1986	167.9		399.48	137.3	148.7
1987	180.9		427.98	147.5	162.6
1988	196.8		454.48	158.6	174.5
1989	214.9		487.30	171.6	187.3
1990	231.9		520.95	184.3	202.3
1991	243.7		555.40	195.1	213.0
1992	245.2		580.75	200.2	217.0
1993	245.7		591.08	202.0	219.2
1994	249.2		608.78	206.5	223.5
1995	255.9		634.00	213.6	230.7
1996	261.2		662.43	220.6	232.9
1997	262.4		688.23	225.4	236.0
1998	264.7		716.65	231.0	236.0
1999	266.9	100.0	743.68	236.3	238.9
2000	269.2		768.55	241.3	244.6
2001	269.9		808.57	247.9	259.3
2002	275.0		853.57	257.1	266.8
2003	286.4		897.40	269.0	275.0
2004	289.0		952.83	278.7	281.4
2005	300.7	116	1008.10	292.4	288.2
2006		120	1032.00	300.9	299.2
2007		124	1091.03	314.5	304.4

Continued

Index estimated by the author based on data published by the Australian Bureau of Statistics—cont'd

Financial year ending June	Materials price index	Producer price index	Average weekly earnings ($)	Plant cost index	Consumer price index
2008		131	1142.28	330.8	317.4
2009		129	1201.10	336.8	312.5
2010		132.5	1256.70	349.1	321.0
2011		137.0	1315.03	363.1	332.6
2012		138.5	1353.66	370.4	336.6
2013		140.2	1422.70	382.1	344.7
2014		141.7	1477.00	391.3	355.0
2015		143.0	1500.50	396.2	360.3
2016		145.1	1516.00	401.1	363.9
2017		148.3	1543.80	409.2	370.8
2018		150.5	1589.85	418.3	378.6
2019		153.5	1637.55	428.8	384.7

Comments: *The Australian plant cost index was initially estimated based on a 50% weighting of the price index for materials used in buildings other than houses (reported monthly) and a 50% weighting of average weekly earnings for ordinary full-time adults (reported quarterly) averaged for the stipulated year. The ABS ceased publishing the price index for buildings other than houses and published a producer price index (providing similar data) from around 2000. The consumer price index is included for comparison.*

Average weekly earnings are in $ Australian.

APPENDIX

US inflation indices for plant capital cost and cost of living 4

Year	Plant cost inflation indices		Consumer price index
	United States	United States	United States
	CE plant cost index	Nelson Farrar inflation index	CPI Annual average
1959	100		29.1
1960	102	228	29.6
1965	104	261	31.5
1970	126	365	38.8
1975	182	576	53.8
1980	261	823	82.4
1985	325	1074	107.6
1990	357.6	1226	130.7
1995	381.1	1392	152.4
2000	394.1	1543	172.2
2001	394.3	1580	177.1
2002	395.6	1642	179.9
2003	402	1710	184.0
2004	44.2	1833	188.9
2005	468.2	1894	195.3
2006	499.6	2008	201.6
2007	525.4	2017	207.3
2008	575.4	2251	215.3
2009	521.9	2218	214.5
2010	550.8	2338	218.1
2011	585.7	2436	224.9
2012	584.6	2465	229.6
2013	567.3	2490	233.0
2014	576.1	2555	236.7
2015	556.8	2550	237.0
2016	541.7	2599	240.0
2017	567.5	2689	244.0
2018	603.1	Not available	251.1
2019	615 approximate	Not available	256 approximate

References

[1] 'Chemical Engineering' Journal for CE Plant Cost Index (www.chemengonline.com).
[2] 'Oil and Gas Journal' for Nelson Farrar Index (www.ogj.com).
[3] US Dept of labour, Bureau of Labour Statistics, Washington, DC for CPI (www.bls.gov/cpi).

APPENDIX

Discount tables

5

The tables give the value of $\frac{1}{(1+i)^t}$ for values of t from 1 year to 20 years and of i from 0.01 (1%) to 0.20 (20%).

Year t	*i*, percentage				
	1	2	3	4	5
1	0.9901	0.9804	0.9709	0.9615	0.9524
2	0.9803	0.9612	0.9426	0.9246	0.9070
3	0.9706	0.9423	0.9151	0.8890	0.8638
4	0.9610	0.9238	0.8885	0.8548	0.8227
5	0.9515	0.9057	0.8626	0.8219	0.7835
6	0.9420	0.8880	0.8375	0.7903	0.7462
7	0.9327	0.8706	0.8131	0.7599	0.7107
8	0.9235	0.8535	0.7894	0.7307	0.6768
9	0.9143	0.8368	0.7664	0.7026	0.6446
10	0.9053	0.8203	0.7441	0.6756	0.6139
11	0.8963	0.8043	0.7224	0.6496	0.5847
12	0.8874	0.7885	0.7014	0.6246	0.5568
13	0.8787	0.7730	0.6810	06006	0.5303
14	0.8700	0.7579	0.6611	0.5775	0.5051
15	0.8613	0.7430	0.6419	0.5553	0.4810
16	0.8528	0.7284	0.6232	0.5339	0.4581
17	0.8444	0.7142	0.6050	0.5134	0.4363
18	0.8360	0.7002	0.5874	0.4936	0.4155
19	0.8277	06864	0.5703	0.4746	0.3957
20	0.8195	0.6730	0.5537	0.4564	0.3769

Year *t*	*i*, percentage				
	6	7	8	9	10
1	0.9434	0.9346	0.9259	0.9174	0.9091
2	0.8900	0.8734	0.8573	0.8417	0.8264
3	0.8396	0.8163	0.7938	0.7722	0.7513
4	0.7921	0.7629	0.7350	0.7084	0.6830
5	0.7473	0.7130	0.6806	0.6499	0.6209
6	0.7050	0.6663	0.6302	0.5963	0.5645
7	0.6651	0.6227	0.5835	0.5470	0.5132
8	0.6274	0.5820	0.5403	05019	0.4665
9	0.5919	0.5439	0.5002	0.4604	0.4241
10	0.5584	0.5083	0.4632	0.4224	0.3855
11	0.5268	0.4751	0.4289	0.3875	0.3505
12	0.4970	0.4440	0.3971	0.3555	0.3186
13	0.4688	0.4150	0.3677	0.3262	0.2897
14	0.4423	0.3878	0.3405	0.2992	0.2633
15	0.4173	0.3624	0.3152	0.2745	0.2394
16	0.3936	0.3387	0.2919	0.2519	0.2176
17	0.3714	0.3166	0.2703	0.2311	0.1978
18	0.3503	0.2959	0.2502	0.2120	0.1799
19	0.3305	0.2765	0.2317	0.1945	0.1635
20	0.3118	0.2584	0.2145	0.1784	0.1486

Year *t*	*i*, percentage				
	11	12	13	14	15
1	0.9009	0.8929	0.8850	0.8772	0.8696
2	0.8116	0.7972	0.7831	0.7695	0.7561
3	0.7312	0.7118	0.6931	0.6750	0.6575
4	0.6587	0.6355	0.6133	0.5921	0.5718
5	0.5935	0.5674	0.5428	0.5194	0.4972
6	0.5346	0.5066	0.4803	0.4556	0.4323
7	0.4817	0.4523	0.4251	0.3996	0.3759
8	0.4339	0.4039	0.3762	0.3506	0.3269
9	0.3909	0.3606	0.3329	0.3075	0.2843
10	0.3522	0.3220	0.2946	0.2697	0.2472
11	0.3173	0.2875	0.2607	0.2366	0.2149
12	0.2858	0.2567	0.2307	0.2076	0.1869
13	0.2575	0.2292	0.2042	0.1821	0.1625

Continued

Year *t*	*i*, percentage				
	11	**12**	**13**	**14**	**15**
14	0.2320	0.2046	0.1807	0.1597	0.1413
15	0.2090	0.1827	0.1599	0.1401	0.1229
16	0.1883	0.1631	0.1415	0.1229	0.1069
17	0.1696	0.1456	0.1252	0.1078	0.0929
18	0.1528	0.1300	0.1108	0.0946	0.0808
19	0.1377	0.1161	0.0981	0.0829	0.0703
20	0.1240	0.1037	0.0868	0.0728	0.0611

Year *t*	*i*, percentage				
	16	**17**	**18**	**19**	**20**
1	0.8621	0.8547	0.8475	0.8403	0.8333
2	0.7432	0.7305	0.7182	0.7062	0.6944
3	0.6407	0.6244	0.6086	0.5934	0.5787
4	0.5523	0.5337	0.5158	0.4987	0.4823
5	0.4761	0.4561	0.4371	0.4190	0.4019
6	0.4104	0.3898	0.3704	0.3521	0.3349
7	0.3538	0.3332	0.3139	0.2959	0.2791
8	0.3050	0.2848	0.2660	0.2487	0.2326
9	0.2630	0.2434	0.2255	0.2090	0.1938
10	0.2267	0.2080	0.1911	0.1756	0.1615
11	0.1954	0.1778	0.1619	0.1476	0.1346
12	0.1685	0.1520	0.1372	0.1240	0.1122
13	0.1452	0.1299	0.1163	0.1042	0.0935
14	0.1252	0.1110	0.0985	0.0876	0.0779
15	0.1079	0.0949	0.0835	0.0736	0.0649
16	0.0930	0.0811	0.0708	0.0618	0.0541
17	0.0802	0.0693	0.0600	0.0520	0.0451
18	0.0691	0.0592	0.0508	0.0437	0.0376
19	0.0596	0.0506	0.0431	0.0367	0.0313
20	0.0514	0.0433	0.0365	0.0308	0.0261

Index

Note: Page numbers followed by *f* indicate figures, *t* indicate tables, *b* indicate boxes, and *ge* indicate glossary terms.

9780128194669